Lecture Notes in Mathematics 1534

Cornelius Greither

Cyclic Galois Extensions of Commutative Rings

Springer-Verlag
Berlin Heidelberg New York
London Paris Tokyo
Hong Kong Barcelona
Budapest

Author

Cornelius Greither
Mathematisches Institut
der Universität München
Theresienstr. 39
W-8000 München 2, Germany

Mathematics Subject Classification (1991): 11R18, 11R23, 11R33, 11S15, 13B05, 13B15, 14E20

ISBN 3-540-56350-4 Springer-Verlag Berlin Heidelberg New York
ISBN 0-387-56350-4 Springer-Verlag New York Berlin Heidelberg

Printed in Germany

Typesetting: Camera ready by author
46/3140-543210 - Printed on acid-free paper

CONTENTS

Für Liane und für Margarete.

INTRODUCTION

The subject of these notes is a part of commutative algebra, and is also closely related to certain topics in algebraic number theory and algebraic geometry. The basic problems in Galois theory of commutative rings are the following: What is the correct definition of a Galois extension? What are their general properties (in particular, in comparison with the field case)? And the most fruitful question in our opinion: Given a commutative ring R and a finite abelian group G, is there any possibility of describing *all* Galois extensions of R with group G?

These questions will be dealt with in considerable generality. In later chapters, we shall then apply the results in number-theoretical and geometrical situations, which means that we consider more special commutative rings: rings of integers and rings of functions. Now algebraic number theory as well as algebraic geometry have their own refined methods to deal with Galois extensions: in number theory one should name class field theory for instance. Thus, the methods of the general theory for Galois extensions of rings are always in competition with the more special methods of the discipline where they are applied. It is hoped the reader will get a feeling that the general methods sometimes also lead to new results and provide an interesting approach to old ones.

Let us briefly review the development of the subject. Hasse (1949) seems to have been the first to consider the totality of G-Galois extensions L of a given number field K. He realized that for finite abelian G this set admits a natural abelian group structure, *if* one also admits certain "degenerate" extensions L/K which are not fields. For example, the neutral element of this group is the direct product of copies of K, with index set G. This constitutes the first fundamental idea. The second idea, initiated by Auslander and Goldman (1960) and then brought to perfection by Chase, Harrison, and Rosenberg (1965), is to admit base rings R instead of fields. It is not so obvious what the definition of a G-Galois extension S/R of commutative rings should be, but once one has a good definition (by the way, all good definitions turn out to be equivalent), then one also obtains nice functoriality properties, stability under base change for instance, and the theory runs almost as smoothly as for fields. Harrison (1965) put the two ideas together and defined, for G finite abelian, the *group* of all G-Galois extensions of a given commutative ring R modulo G-isomorphism. This group is now called the *Harrison group*, and we denote it by $\mathrm{H}(R,G)$. Building on the general theory of Chase, Harrison, and Rosenberg, and developing some new tools, we calculate in these notes the group $\mathrm{H}(R,G)$ in a fairly general setting.

The principal link between this theory and number theory is the study of ramification. Suppose L is a G-Galois extension of the number field K, Σ a set of finite places of K, and $R = \mathcal{O}_{K,\Sigma}$ the ring of Σ-integers in K. Then the integral closure S of R in L is with the given G-action a G-Galois extension of R if and only if L/K is at most ramified in places which belong to Σ. In most applications, Σ will be the set of places over p. The reason for this choice will become apparent when we discuss $\mathbb{Z}_p$-extensions below.

We now discuss the contents of these notes in a little more detail.

After a summary of Galois theory of rings in Chap. 0, which also explains the connection with number theory, and $\mathbb{Z}_p$-extensions, we develop in Chap. I a *structure theory* for Galois extensions with cyclic group $G = C_{p^n}$ of order p^n, under the hypothesis that $p^{-1} \in R$ and p is an odd prime number. For technical reasons, we also suppose that R has no nontrivial idempotents. Since the Harrison group $H(R,G)$ is functorial in both arguments, and preserves products in the right argument, this also gives a structure theory for the case G finite abelian, $|G|^{-1} \in R$.

The basic idea is simple. If R contains a primitive p^n-th root of unity ζ_n (this notion has to be defined, of course), and $p^{-1} \in R$, then Kummer theory is available for C_{p^n}-extensions of R. The statements of Kummer theory are, however, more complicated than in the field case: it is no longer true that every C_{p^n}-extension S/R can be gotten by "extracting the p^n-th root of a unit of R", but the obstruction is under control. The procedure is now to adjoin ζ_n to R somehow (it is a lot of work to make this precise), use Kummer theory for the ring S_n obtained in this way, and descend again. Here a very important concept makes its appearance. A G-Galois extension S/R is defined to have *normal basis*, if S has an R-basis of the form $\{\gamma(x) \mid \gamma \in G\}$ for some $x \in S$. Fo $G = C_{p^n}$, the extensions with normal basis make up a *subgroup* $NB(R,C_{p^n})$ of $H(R,C_{p^n})$. In Chap. I we prove rather precise results on the structure of $NB(R,C_{p^n})$, and of $H(R,C_{p^n})/NB(R,C_{p^n})$. In the field case, the latter group is trivial, but not in general. Kersten and Michaliček (1988) were the first to prove results for $NB(R,C_{p^n})$. Our result says that $NB(R,C_{p^n})$ is "almost" isomorphic to an explicitly given subgroup of $S_n^*/(p^n$-th powers), and $H(R,C_{p^n})/NB(R,C_{p^n})$ is isomorphic to an explicitly given subgroup of the Picard group of S_n. The description of $NB(R,C_{p^n})$ is basic for the calculations in Chap. III and V.

In Chap. II we treat corestriction and a result of type "Hilbert 90". This amounts to the following: We get another description of $NB(R,C_{p^n})$, this time as a *factor* group of $S_n^*/(p^n$-th powers). This is sometimes more practical, as witnessed by the *lifting theorems* which conclude Chap. II: If I is an ideal of R, contained in the Jacobson radical of R, then every C_{p^n}-extension S of R/I with normal basis is of the form $S \cong T/IT$, $T \in NB(R,C_{p^n})$.

In Chap. III we set out to calculate the order of $\mathrm{NB}(R, C_{p^n})$, where now $R = \mathcal{O}_K[p^{-1}]$, K a number field. Although one almost never knows the groups S_n^* explicitly, which are closely related to the group of units in the ring of integers of $K(\zeta_n)$, one can nevertheless do the calculation one wants, by dint of some tricks involving a little cohomology of groups. All this is presented in a quite elementary way. We demonstrate the strength of the method by deducing the Galois theory of finite fields, and a piece of local class field theory. The main result for number fields K is that with R as above, and n not "too small", the order of $\mathrm{NB}(R, C_{p^n})$ equals $\mathrm{const}\cdot p^{(1+r_2)n}$, where r_2 is half the number of nonreal embeddings $K \to \mathbb{C}$ as usual.

The goal of Chap. IV is to get an understanding, how far the subgroup $\mathrm{NB}(R, C_{p^n})$ differs from $\mathrm{H}(R, C_{p^n})$, and a similar question for $\mathbb{Z}_p$ in the place of C_{p^n}. Here $\mathrm{H}(R, \mathbb{Z}_p)$ is the group of $\mathbb{Z}_p$-extensions of R. A $\mathbb{Z}_p$-extension is basically a tower of C_{p^n}-extensions, $n \to \infty$. It is known that all $\mathbb{Z}_p$-extensions of K are unramified outside p, and hence already a $\mathbb{Z}_p$-extensions of R, which justifies the choice of the ring R.

We prove in IV §2: $\mathrm{NB}(R, \mathbb{Z}_p) \cong \mathbb{Z}_p^{1+r_2}$. This was previously proved in a special case by Kersten and Michaliček (1989). The result is what one expects from the formula for $|\mathrm{NB}(R, C_{p^n})|$, but the passage to the limit presents some subtleties. The index $q_n = [\mathrm{H}(R, C_{p^n}):\mathrm{NB}(R, C_{p^n})]$ is studied in some detail, and we show that q_n either goes to infinity or is eventually constant for $n \to \infty$. The first case conjecturally never happens: we prove that this case obtains if and only the famous Leopoldt conjecture fails for K and p. Another way of saying this is as follows: $\mathrm{NB}(R, \mathbb{Z}_p)$ has finite index in $\mathrm{H}(R, \mathbb{Z}_p)$ if and only if the Leopoldt conjecture is true for K and p. We give results about the actual value of that index; in particular, it can be different from 1.

Apart from adjoining roots of unity, there is so far only other explicit way of generating large abelian extensions of a number field K, namely, adjoining torsion points on abelian varieties with complex multiplication. We show in IV §5 that $\mathbb{Z}_p$-extensions obtained in that way tend to have normal bases over $R = \mathcal{O}_K[p^{-1}]$, and a weak converse to this statement. These results are in tune with the much more explicit results of Cassou-Noguès and Taylor (1986) for elliptic curves.

There is a change of scenario in Chap. V. There we consider function fields of varieties over number fields. Such function fields are also called *absolutely finitely generated fields over* $\mathbb{Q}$. After some prerequisites from algebraic geometry, we show a relative finiteness result on C_{p^n}-Galois coverings of such varieties, which is similar to results of Katz and Lang (1981), and we prove that *all* $\mathbb{Z}_p$-extensions of an absolutely finitely generated field K already come from the greatest number field k contained in K. In other words: for number fields k one does not know how many independent $\mathbb{Z}_p$-extensions k has, unless Leopoldt's conjecture is known to be true for K and p, but in a geometric situation, no new $\mathbb{Z}_p$-extensions arise.

The last chapter (Chap. VI) proposes a structure theory for Galois extensions with group C_{p^n}, in case the ground ring R contains a primitive p^n-th root of unity ζ_n but not necessarily $p^{-1} \in R$. It is assumed, however, that p does not divide zero in R. Even though Kummer theory fails for R, we may still associate to many C_{p^n}-extensions S/R a class $\varphi_n(S) = [u]$ in R^* mod p^n-th powers. If R is normal, S will be the integral closure of R in $R[p^{-1}, \sqrt[p^n]{u}]$. The main question is: Which units $u \in R^*$ may occur here? In §2 we essentially perform a reduction to the case R p-adically complete. Taking up a paper of Hasse (1936), we then answer our question by using so-called Artin–Hasse exponentials. It turns out that the admissible values u are precisely the values of certain universal polynomials, with parameters running over R. Reduction mod p also plays an essential role, and for this reason we have to review Galois theory in characteristic p in §1. In the final §6 the descent technique of Chap. I comes back into play. In §4–5 a "generic" C_{p^n}-extension of a certain universal p-complete ring containing ζ_n (but not p^{-1}) was constructed, and we are now able to see in detail how this extension descends down to a similar ground ring without ζ_n, to wit: the p-adic completion of $\mathbb{Z}[X]$. This extension is, roughly speaking, a prototype of C_{p^n}-extensions of p-adically complete rings. All this is in principle calculable.

Most chapters begin with a short overview of their contents. Cross references are indicated in the usual style: the chapters are numbered **0**, I, II, ..., VI, and a reference number not containing **0** or a Roman numeral means a reference within the same chapter. *All rings are supposed commutative* (except, occasionally, an endomorphism ring), and with unity. Other conventions are stated where needed.

Earlier versions of certain parts of these notes are contained in the journal articles Greither (1989), (1991).

It is my pleasurable duty to thank my colleagues who have helped to improve the contents of these notes. Ina Kersten has influenced the presentation of earlier versions in many ways and provided valuable information. Also, the helpful and detailed remarks of several referees are appreciated; I like to think that their suggestions have resulted in a better organization of the notes. Finally, I am grateful for written and oral communications to S. Ullom, G. Malle, G. Janelidze, and T. Nguyen Quang Do.

CHAPTER 0

Galois theory of commutative rings

§1 Definitions and basic properties

The study of Galois extensions of commutative rings was initiated by Auslander and Goldman (1960) and developed by Chase, Harrison, and Rosenberg (1965). In this section we shall try to present the basics of this theory. Occasionally we refer to the paper of Chase, Harrison, and Rosenberg for a proof. Almost everything we say in this section is can be found there, or in the companion paper Harrison (1965), sometimes with proofs which differ from ours.

Let G be a finite group, $K \subset L$ a field extension. Then, as everybody agrees, L/K is a Galois extension with group G if and only if:

G is a subgroup of $\mathrm{Aut}(L/K)$, the group of automorphisms of L which fix all elements of K; and

$K = L^G$, the field of all elements of L which are fixed by every automorphism in G.

A literal translation of this definition would result in a too weak definition in the framework of commutative rings, for many reasons. Let us not pursue this, but rather point out two alternative definitions of "Galois extension" in the field case which turn out to generalize well, and which indeed give equivalent generalizations. Thus, we will have found the "correct" notion of a Galois extension of commutative rings. Suppose that G is a finite group which acts on L by automorphisms which fix all elements of K. We thus have a group homomorphism $G \to \mathrm{Aut}(L/K)$.

Definition 1.1. The K-algebra $L*G$ is the L-vectorspace $\bigoplus_{\sigma\in G} Lu_\sigma$ (the u_σ are just formal symbols), with multiplication given by $(\lambda u_\sigma)(\mu u_\tau) = \lambda\cdot\sigma(\mu)\cdot u_{\sigma\tau}$ $(\lambda,\mu\in L)$. The map j: $L*G \to \mathrm{End}_K(L)$ is given by

$$j(\lambda u_\sigma) = (\mu \longmapsto \lambda\cdot\sigma(\mu)) \in \mathrm{End}_K(L).$$

Proposition 1.2. *j is a well-defined K-algebra homomorphism, which is bijective iff G is embedded in* $\mathrm{Aut}(L/K)$ *and L/K is a G-Galois extension.*

Proof. The first statement is easy to check. Assume $G \subset \mathrm{Aut}(L/K)$ and L/K is G-Galois. Then by Dedekind's Lemma the elements σ of G are L-left linearly independent in $\mathrm{End}_K(L)$, hence j is a monomorphism. Since $\dim_K(L*G) = [L:K]^2 = \dim_K \mathrm{End}_K(L)$, j is bijective.

If $G \to \mathrm{Aut}(L/K)$ is not injective, then there exist $\sigma \neq \tau$ in G with $j(\sigma) = j(\tau)$, i.e. j cannot be monic. If G embeds into $\mathrm{Aut}(L/K)$ but L/K fails to be G-Galois, then there exists $x \in L \setminus K$ fixed under G. A short calculation shows then that $l_x =$ (left multiplication by x) commutes with $\mathrm{Im}(j) \subset \mathrm{End}_K(L)$. If j were surjective, we would have l_x in the center of $\mathrm{End}_K(L)$, i.e. $x \in K$, contradiction.

Definition 1.3. The K-algebra $L^{(G)}$ is defined to be the set of all maps $G \to L$, endowed with the obvious addition and multiplication. (Note that L^G, without brackets, denotes a fixed field.) Let $h\colon L \otimes_K L \longrightarrow L^{(G)}$ be defined by $h(x \otimes y) = (x \cdot \sigma(y))_{\sigma \in G}$.

Proposition 1.4. *The map h is a L-algebra homomorphism (here L operates on the left factor of $L \otimes_K L$), and h is bijective iff G embeds into* $\mathrm{End}_K(L)$ *and L/K is G-Galois.*

Proof. The first statement is obvious. Pick a K-basis $y_1, \ldots, y_n$ of L. Then $1 \otimes y_1, \ldots, 1 \otimes y_n$ is an L-basis of $L \otimes_K L$. Thus we see that h is bijective iff the matrix $(\sigma(y_i))_{\sigma \in G, 1 \leq i \leq n}$ has full rank (note that this is indeed a square matrix). The latter condition says that the images of all $\sigma \in G$ are L-left linearly independent in $\mathrm{End}_K(L)$, or (what is the same) that the map j of 1.1 is injective. Hence 1.4 follows from 1.2.

Motivated by these descriptions of Galois field extensions, we define for any finite group G:

Definition 1.5. An extension $R \subset S$ of commutative rings is a *G-Galois extension*, if G is a subgroup of $\mathrm{Aut}(S/R) = \{\varphi\colon S \to S \mid \varphi\ R\text{-algebra automorphism}\}$, such that $R = S^G$ (fixed ring under G), and the map $h\colon S \otimes_R S \longrightarrow S^{(G)}$, $h(x \otimes y) = (x\sigma(y))_{\sigma \in G}$ exactly as in 1.3, is bijective, or (what is the same) an S-algebra isomorphism.

Examples: a) Galois extensions of fields are obviously a special case.

b) For any commutative ring R we have the *trivial G-extension* $S = R^{(G)}$ which is defined as follows: The algebra $R^{(G)}$ is again just $\mathrm{Map}(G, R)$ with the canonical R-algebra structure, and the action of G is given by index shift:

$$\sigma\big((x_\tau)_{\tau \in G}\big) = (x_{\tau\sigma})_{\tau \in G} \quad \text{for } \sigma \in G,\ (x_\tau)_{\tau \in G} \in R^{(G)}.$$

It is an easy exercise to prove that in this case indeed $S^G = R$ and h is bijective. We shall see more examples below.

There exist plenty of other definitions, or rather characterizations, of G-Galois extensions of commutative rings. Some of them are listed in the next theorem:

Theorem 1.6. [Chase-Harrison-Rosenberg (1965), Thm. 1.3]: *Let $R \subset S$ be commutative rings, $G \subset$* $\mathrm{Aut}(S/R)$ *a finite subgroup such that $S^G = R$. Then the following conditions are equivalent*:

(*i*) *S/R is G-Galois (i.e. per def.: $h\colon S \otimes_R S \longrightarrow S^{(G)}$ is bijective)*;

(*ii*) *$h\colon S \otimes_R S \longrightarrow S^{(G)}$ is surjective*;

(iii) S is a finitely generated projective R-module, and the map j: $S*G \to \mathrm{End}_R(S)$ (defined as in 1.1) is bijective;

(iv) For any $\sigma \in G\backslash\{e_G\}$ and any maximal ideal $M \subset S$, there exists $y \in S$ with $\sigma(y) - y$ not in M.

Proof. Let us first reformulate condition *(ii)*. One sees easily that *h* is compatible with the *G*-action, where *G* acts naturally on the second factor of $S \otimes_R S$, and by index shift on $S^{(G)}$, exactly as in example b) above. Therefore *h* is surjective iff the element (1, 0, . . ., 0) is in Im(*h*) (the 1 is at position e_G). Letting $\sum x_i \otimes y_i$ be a preimage of (1, 0, . . ., 0) under *h*, we get the following reformulation of *(ii)*:

(ii') There exist $n \in \mathbb{N}$ and $x_1,\ldots,x_n$, $y_1,\ldots,y_n \in S$ such that $\sum_{i=1}^n x_i\sigma(y_i)$ is 1 or 0, according to whether $\sigma = e_G$ or $\sigma \neq e_G$. (We may write $\sum_{i=1}^n x_i\sigma(y_i) = \delta_{\sigma,e}$.)

(i) ⇒ *(ii)*: This is trivial.

(ii') ⇒ *(iii)*: We first show that ${}_RS$ is finitely generated projective. Define the *trace* tr: $S \to R$ by $tr(y) = \sum_{\sigma\in G}\sigma(y)$. ($tr$ is well-defined since $S^G = R$, and *R*-linear since all $\sigma \in G$ are *R*-linear.) Let φ_i: $S \to R$ be defined by $\varphi_i(z) = tr(zy_i)$, $z \in S$. Then the formula of *(ii')* implies by direct calculation: $z = \sum_i \varphi_i(z)\cdot y_i$ for all $z \in S$, i.e. the pairs (x_i, φ_i) are a dual basis for ${}_RS$, which is hence finitely generated projective.

Now we may, by localization, assume hat *S* is even finitely generated free over *R*, with basis $x_1',\ldots,x_n'$, say. We may then assume that the x_i in condition *(ii')* are just the x_i', because every element of $S \otimes S$ can be written in the form $\sum x_i' \otimes y_i'$, and it does not matter just how we write a preimage of (1,0,...0) under *h*. Let us therefore omit the ' again. From the calculation just performed we get $x_j = \sum_i \varphi_i(x_j)\cdot x_i$, hence by definition of φ_i, and since the x_i are a basis, $tr(x_jy_i) = \delta_{ij}$. As in the field case, bijectivity of *j* is equivalent to invertibility of the matrix $A = (\sigma(x_i))_{\sigma,i}$. One calculates as follows: Let $B = (\tau(y_j))_{j,\tau}$. Then $AB = (\delta_{\sigma,\tau})$ = unit matrix (use *(ii')*), and $BA = (tr(x_jy_i))_{ji}$ = unit matrix. Hence *A* is invertible, and *j* is bijective.

(iii) ⇒ *(i)*: Since *S* is finitely generated projective over *R*, we may again assume that *S* is free over *R*, with basis $x_1,\ldots,x_n$. As in the last paragraph, *j* is bijective iff the matrix $A = (\sigma(x_i))_{\sigma,i}$ is invertible. As in the field case, this is again equivalent to the bijectivity of *h*.

(ii') ⇒ *(iv)*: Just suppose $\sigma \neq e_G$ ($= id$), and $\sigma(y) - y \in M$ for all $y \in S$. Then $1 = \sum_i x_i(y_i - \sigma(y_i)) \in M$, contradiction.

(iv) ⇒ *(ii')*: We first construct a solution of the formula in *(ii')* for a single $\sigma \neq e_G$. By *(iv)*, the ideal of *S* generated by all $y - \sigma(y)$ is contained in no maximal ideal, hence is equal to *S*. One finds hence $n_\sigma \in \mathbb{N}$ and $x_1^{(\sigma)},\ldots,x_{n_\sigma}^{(\sigma)}$, $y_1^{(\sigma)},\ldots,y_{n_\sigma}^{(\sigma)} \in S$ with $\sum_i x_i^{(\sigma)}\cdot\left(y_i^{(\sigma)} - \sigma(y_i^{(\sigma)})\right) = 1$. Now one lets $x_0 = \sum_{i=1}^{n_\sigma} x_i^{(\sigma)}\cdot\sigma(y^{(\sigma)})$ and $y_0 = -1$.

We then get (summation from 0 to n_σ): $\sum x_i^{(\sigma)} \cdot y_i^{(\sigma)} = 1$ and $\sum x_i^{(\sigma)} \cdot \sigma(y_i^{(\sigma)}) = 0$. Now one shuffles together these solutions for individual σ to a solution for all σ as follows: Let I be the index set $\prod_{\sigma \in G \setminus e} \{0,\dots,n_\sigma\}$; for each $\imath \in I$, let $x_\imath$ be the product of all $x_{\imath(\sigma)}^{(\sigma)}$ with $\sigma \neq e$, and $y_\imath$ similarly. One can then check that indeed for all $\sigma \in G$: $\sum_{\imath \in I} x_\imath \cdot \sigma(y_\imath)$ is equal to $\delta_{\sigma,e}$, q.e.d.

In our opinion, it is instructive to use the theory of faithfully flat descent already at this early stage of Galois theory of rings. To this end, recall that an R-module M is *faithfully flat* if M is flat, and $M/PM \neq 0$ for each maximal ideal P of R. It is another characterization of faithful flatness that the functor $M \otimes_R -$ preserves and detects short exact sequences of R-modules. One has the following easy results:

Proposition 1.7. [Knus-Ojanguren (1974), Bourbaki Alg. comm. I §3] *Let M be a faithfully flat R-module, and φ: $A \to B$ a homomorphism of R-modules. Then φ is an isomorphism iff $M \otimes_R \varphi$: $M \otimes_R A \to M \otimes_R B$ is an isomorphism. The statement remains correct, if the word "isomorphism" is replaced by "monomorphism", or by "epimorphism".*

This simple result already has applications. Suppose T is an R-algebra which is a faithfully flat R-module, and suppose S is a ring extension of R such that the finite group G acts on S by R-automorphisms. One can then state

Proposition 1.8. *Under these hypotheses, S/R is a G-Galois extension if $T \otimes_R S$ is a G-Galois extension over T.*

Proof. We may consider $T = T \otimes_R R$ as a subalgebra of $T \otimes_R S$, since T is flat. We use the defining property of "G-Galois" and check that the map

$$h_T: (T \otimes_R S) \otimes_T (T \otimes_R S) \longrightarrow (T \otimes_R S)^{(G)}$$

associated in Def. 1.5. with the extension $T \subset T \otimes_R S$ is, up to canonical isomorphism, just $T \otimes h$ (where h: $S \otimes_R S \longrightarrow S^{(G)}$ is the map of Def. 1.5. for the extension $R \subset S$). By 1.7, if $T \otimes h$ is an isomorphism, then so is h. We still have to show that $S^G = R$, i.e. the canonical map ι: $R \to S^G$ is onto. But it follows from the flatness of T that $(T \otimes_R S)^G \cong T \otimes_R S^G$, hence $T \otimes \iota$ is onto. By 1.7, we are done.

The converse of 1.8 is "more than true", in the sense that base change always preserves G-Galois extensions (not only faithfully flat base change). We will see this a little later.

Lemma 1.9. *Any G-Galois extension S/R is faithfully flat over R.*

Proof. Flatness is clear since S/R is projective by Thm. 1.6 (*iii*). Pick a maximal ideal P of R; we need $S/PS \neq 0$. By Nakayama, it suffices to see $S_P \neq 0$. But $R \subset S$, and localization preserves monomorphisms, so we are done.

This lemma suggest to try out S in the role of T; the result is strikingly simple, but we first need to define morphisms of G-Galois extensions:

Definition. If S and S' are two G-Galois extensions, then a *morphism* $\varphi: S \to S'$ is a G-equivariant R-algebra homomorphism from S to S'. (G-equivariance means of course: $\varphi(\sigma x) = \sigma\varphi(x)$ for all $\sigma \in G$, $x \in S$.) The G-Galois extension S/R is called *trivial*, if it is isomorphic to the trivial extension $R^{(G)}/R$.

Remark. It is obvious that we obtain a *category* GAL(R,G) of G-Galois extensions of a given ring R.

Now we can see that "base-extending any Galois extension with itself gives a trivial extension". More precisely: Let S/R be a G-Galois extension, let $T = S$, and consider the ring extension $T\otimes_R S/T$. Since $T = S$, it is now easy to check that the isomorphism h: $S\otimes_R S \to S^{(G)}$ gives an isomorphism of G-Galois extensions h: $T\otimes_R S \to T^{(G)}$. Recall that G operates naturally on the second factor S, and by index shift on $S^{(G)}$. We now can prove a result on the trace:

Lemma 1.10. *Let S/R be G-Galois, and tr: $S \longrightarrow R$ the trace (see proof (ii') $\Rightarrow$ (iii) of 1.6). Then:*

a) tr: $S \to R$ is surjective

b) The R-submodule R of S is a direct summand of S.

Proof. a) By the previous remarks, $S\otimes S/S\otimes R$ ($= S$) is isomorphic to the trivial extension of S. One has a commutative diagram

$$\begin{array}{ccc} S\otimes_R S & \xrightarrow{\cong} & S^{(G)} \\ {\scriptstyle S\otimes tr}\downarrow & & \downarrow{\scriptstyle tr_S} \\ S\otimes_R R & \xrightarrow{\cong} & S \end{array},$$

where tr_S is the trace associated to the extension $S^{(G)}/S$. S is embedded diagonally in $S^{(G)}$, and one sees from the way G acts on $S^{(G)}$ that $tr_S(x,0,...,0) = (x,x,...,x) =$ diag(x) for all $x \in S$. Hence tr_S is onto; by 1.7, tr is onto.

b) Pick $c \in S$ with $tr(c) = 1$, and let f: $S \to R$ be defined by $f(x) = tr(cx)$. Then f is an R-linear section of the inclusion $R \subset S$, so R is a direct summand of S.

Now we can show:

Lemma 1.11. *Let S/R be G-Galois, and T any R-algebra. Then $T\otimes_R S/S$ is again a G-Galois extension.*

Proof. Write S_T for $T\otimes_R$ S.We want three things: T embeds in S_T, $S_T{}^G = T$, and h_T: $S_T \otimes_T S_T \longrightarrow S_T{}^{(G)}$ is an isomorphism. The last condition is the easiest to see, since we know already that h_T is (up to canonical isomorphism) just $T\otimes h$, and h is an isomorphism by hypothesis. Since R splits of in S, the map $T \to S_T$ also splits, in particular T is a subring of S_T. To see the second condition, we argue as in

Chase–Harrison–Rosenberg (1965): Pick $c \in S$ with $tr(c) = 1$ (Lemma 1.10). and let $y \in S_T$ be fixed under G. Then $y = (T \otimes tr)(1 \otimes c) \cdot y = \sum_\sigma (1 \otimes \sigma(c)) \cdot y = \sum_\sigma (T \otimes \sigma)((1 \otimes c) \cdot y) = (T \otimes tr)((1 \otimes c) \cdot y) \in \mathrm{Im}(T \otimes tr) = T \otimes R = T$, q.e.d.

As another example of this descent technique, we show the following important fact:

Proposition 1.12. *Let S/R and S'/R be G-Galois. Then every morphism φ: $S \to S'$ of G-Galois extensions is an isomorphism.*

Proof. There exists a faithfully flat R-algebra T such that both S_T ($= T \otimes_R S$) and S'_T are trivial G-extensions of T. (Trivial G-extensions are obviously preserved by arbitrary base change. Hence one can for example take $T = S \otimes_R S'$, since base extension with S (resp. S') trivializes S/R (resp. S'/R).) It is obvious that $T \otimes \varphi$ is a morphism from S_T to S'_T. We may now suppose, by virtue of 1.7, that $T = R$ (fresh notation), $S = S' = R^{(G)}$. Moreover it is harmless to suppose R local. Let now $e^\sigma \in R^{(G)}$ be the element with 1 in position σ and 0 elsewhere ($\sigma \in G$). These e^σ, $\sigma \in G$, are a complete set of irreducible idempotents of $R^{(G)}$, and they are permuted by G in an obvious fashion. In particular, G permutes the e^σ transitively. Getting back to our morphism φ, we now see that the $\varphi(e^\sigma)$ are pairwise orthogonal idempotents with sum 1. If any of them is zero, then all are zero since φ is G-equivariant, so no $\varphi(e^\sigma)$ is zero. Therefore φ must simply permute the e^σ, which implies immediately that φ is an isomorphism.

§ 2 The main theorem of Galois theory

We fix a finite group G and a G-Galois extension S/R of (commutative) rings. Can one find a bijection between subgroups $H \subset G$ and R-subalgebras $U \subset S$? Certainly this problem is not well posed if we admit *all* subalgebras. (Already for $R = \mathbb{Z}$ and $|G| = 2$, the trivial G-extension $S = \mathbb{Z} \times \mathbb{Z}$ has infinitely many subalgebras.) The correct condition to impose on subalgebras is *separability*, an important concept in itself. One may found the whole theory on this concept, which we avoided for the sake of simplicity; we shall use separable algebras practically only in Chapter 0, and as little as possible. Let us just recall the definition and refer the interested reader to DeMeyer–Ingraham (1971). We remind the reader that all rings are supposed commutative.

Definition. An R-algebra S is called *separable* if S is projective as a module over $S \otimes_R S$ (the structure is $(s \otimes t)y = syt$ for $y \in S$, $s \otimes t \in S \otimes S$). If one admits non-commutative algebras S, one has to take $S \otimes S^{opp}$ in the place of $S \otimes S$.

Example. If $R \subset S$ is a field extension of finite degree, then S is a separable R-algebra iff the extension S/R is separable in the usual sense.

Galois extensions are always separable; more precisely, there is the following extension to Theorem 1.6:

Theorem 2.1. *Let S/R be an extension of rings, G a finite subgroup of* Aut(S/R) *such that $S^G = R$. Then the following are equivalent:*

(*i*) *S/R is G-Galois*

(*ii*) *S is separable over R, and for each nonzero idempotent $e \in S$ and any $\sigma, \tau \in G$ with $\sigma \neq \tau$, there exists $y \in S$ with $e \cdot \sigma(y) \neq e \cdot \tau(y)$. (Note that the last condition is vacuously true if S has no idempotents beside* 0 *and* 1.)

Proof. See Chase-Harrison-Rosenberg (1965), Thm. 1.3. The last condition in (*ii*) is abbreviated to "if $\sigma \neq \tau$, then σ and τ are *strongly distinct*" in loc.cit.

To keep matters simple, let us assume from now on that S is *connected*, i.e. S has no idempotents besides 0 and 1. The first part of the Main Theorem runs as follows:

Theorem 2.2. [Chase-Harrison-Rosenberg (1965)] *Let S/R be a G-Galois extension, $H \subset G$ a subgroup, and let $U = S^H$ be the subalgebra of H-invariant elements. Then:*

(*i*) *U is separable over R*

(*ii*) *S is, in the canonical way, an H-Galois extension of U*

(*iii*) *H is the group of all $\sigma \in G$ which leave U pointwise fixed*

(*iv*) *If H is a normal subgroup of G, then U is, in the canonical way, a G/H-Galois extension of R.*

Proof. We include most of the proof, in order to give the reader a better feeling for the theory. Our argument is mainly the original one (loc.cit.); the changes reflect personal tastes and do not claim to be simplifications. Parts of the proof can be understood without any knowledge about separable algebras.

(*ii*): Choose $x_1,\dots,x_n, y_1,\dots,y_n \in S$ as in (*ii'*) (proof of 1.6). Then, a fortiori, $\sum_i x_i\sigma(y_i) = \delta_{\sigma,id}$ for all $\sigma \in H$. The formula $S^H = U$ holds by definition. Hence S/U is H-Galois by Thm. 1.6.

(*i*): By (*ii*) and Thm. 1.6 (*iii*), S is projective over U, hence $S \otimes_R S$ is projective over $U \otimes_R U$. Recalling the definition of separable algebras, we see from Thm. 2.1 that S is projective over $S \otimes_R S$. Hence, by an easy argument, S is projective over $U \otimes_R U$. But U is a direct summand of S (as a U-module, and hence as a $U \otimes U$-module), by (*ii*) and Lemma 1.10 c). Hence U is projective over $U \otimes_R U$, q.e.d.

(*iii*): We reproduce the direct argument of Chase, Harrison, and Rosenberg. Let $H' = \{\sigma \in G | \sigma \text{ fixes } U \text{ pointwise}\}$. Then $H \subset H'$ and $S^{H'} = S^H = U$. Applying (*ii*) and the definition of Galois extension to U and both of H, H', we obtain that $S \otimes_R S$ is simultaneously isomorphic to $S^{(H)}$ and to $S^{(H')}$, which forces $|H| = |H'|$, and hence $H = H'$.

(*iv*): See loc.cit. p.23. Another approach: Reduce by faithfully flat descent to the case $S = R^{(G)}$ and check directly that S^H is canonically isomorphic to $S^{(G/H)}$. By the way: It is not difficult to prove also (*ii*) by this method.

The converse of this theorem reads as follows for connected ground rings R. Warning: for nonconnected R the statement is more involved, see Chase, Harrison, and Rosenberg (1965).

Theorem 2.3. *Let R, S, and G be as in 2.2; let $U \subset S$ be a separable R-subalgebra. Then there is a subgroup H of G with $U = S^H$, and H is of necessity the group of all $\sigma \in G$ fixing U pointwise.*

For the *proof*, we refer to loc.cit. (The theory of separability is used in an essential way.)

§3 Functoriality, and the Harrison product

In this section we summarize the paper of Harrison (1965). Several proofs are omitted.

We have already seen in §1 that any homomorphism f: $R \to T$ of commutative rings induces a functor "base extension" from the category GAL(R,G) of G-Galois extensions of R to the category GAL(S,G). We now consider the second argument with the aim of establishing functoriality in G, too. For motivation, consider a finite group G and a factor group G/N. Then in the classical case there is just one way to associate a G/N-Galois extension with a given G-Galois extension L/K: just take L^N/K. This works for rings just as well, by Thm. 2.2. It is important, however, to allow general group homomorphisms π: $G \to H$. Before giving the construction, let us briefly mention the case where π is the inclusion of G in H. This case has no counterpart in classical Galois theory; it will turn out that in this case the map π^*: GAL(R,G) $\to$ GAL(R,H) is given by a sort of induction process, as in representation theory, and even if S/R is a G-Galois field extension, $\pi^*(S/R)$ is never a field unless $G = H$. Extreme example: $G = e$, and S is the(!) G-Galois extension R of R. Then π^*R will turn out to be the trivial H-extension of R. Now we present the general result.

Theorem 3.1. *a) Let R be a commutative ring R, $\pi\colon G \to H$ a homomorphism of finite groups. Then there is a canonical functor π^*:* $\mathrm{GAL}(R,G) \to \mathrm{GAL}(R,H)$. *If π happens to be a canonical surjection $G \to G/N$, then $\pi^*(S) = S^N$ as in the above discussion.*

(For the construction of π^*, see the proof of this theorem.)

b) The prescription "$\pi \longmapsto \pi^*$" preserves composition up to canonical isomorphism. In other words: If we let $\mathrm{H}(R,G)$ *be the set of isomorphism classes of G-Galois extensions of R, then* $\mathrm{H}(R,G)$ *is again functorial in G, and the prescription "$\pi \longmapsto \mathrm{H}(R,\pi)$" now preserves composition.*

Definition. The set $\mathrm{H}(R,G)$ just defined is also called the *Harrison set* of R and G.

Proof of Thm. 3.1. We do a) and b) simultaneously. First we define π^*. Let $S \in \mathrm{GAL}(R,G)$. We set

$$\pi^*S = \mathrm{Map}_\pi(H,S) \left(= \{x\colon H \to S \,\middle|\, \forall g \in G,\ h \in H\colon x(\pi(g)h) = g(x(h))\}.\right)$$

The H-action on π^*S is given by $(h'*x)(h) = x(h \cdot h')$ for $x \in \mathrm{Map}_\pi(H,S)$, $h, h' \in H$. The R-algebra structure is defined "component-wise", i.e. by the inclusion of $\mathrm{Map}_\pi(H,S)$ in $\mathrm{Map}(H,S) = S^{(H)}$. (It is immediate that $\mathrm{Map}_\pi(H,S)$ is indeed a subalgebra.)

One sees easily that π^* is a functor from $\mathrm{GAL}(R,G)$ in the category of R-algebras with action of H. It remains to establish:

(*i*) If $\psi\colon H \to J$ is another group homomorphism, then we have a natural isomorphism $\psi^*\left(\pi^*S\right) \cong (\psi\pi)^*S$;

(*ii*) π^*S/R is, with the given H-action, indeed an H-Galois extension.

We do (*i*) first, by exhibiting natural bijections

$$\mathrm{Map}_\psi(J, \mathrm{Map}_\pi(H,S)) \overset{\alpha,\beta}{\rightleftarrows} \mathrm{Map}_{\psi\pi}(J, S).$$

(It is left to the reader to verify that α and β are J-equivariant R-algebra homomorphisms.) Let $\alpha(y) = y(-)(e_H)$ for y in the left hand side, i.e. $\alpha(y)(j) = y(j)(e_H)$ for $j \in J$. Let $\beta(z)(j)(h) = z(\psi(h)j)$ for z in the right hand side, $h \in H$, $j \in J$.

We check α is well-defined, i.e. $\alpha(y) \in \mathrm{Map}_{\psi\pi}(J,S)$: Let $j \in J$, $g \in G$. We calculate:

$$\begin{aligned}
\alpha(y)(\psi\pi(g)\cdot j) &= y(\psi\pi(g)\cdot j)(e_H) \\
&= \left(\pi(g)*y(j)\right)(e_H) \quad \text{(since } y \in \mathrm{Map}_\psi \ldots) \\
&= y(j)(e_H\pi(g)) \quad \text{(def. of } H\text{-action on } \mathrm{Map}_\pi(H,S)) \\
&= g\,(y(j)(e_H)) \quad \text{(since } y(j) \in \mathrm{Map}_\pi \ldots)
\end{aligned}$$

$$= g\,(\alpha(y)(j)), \text{ q.e.d.}$$

$\beta\alpha$ is the identity: Let $y \in \mathrm{Map}_\psi(J, \mathrm{Map}_\pi(H,S))$, $j \in J$, $h \in H$. Then $(\beta\alpha(y))(j)(h) = \alpha(y)(\psi(h)j) = y(\psi(h)\,j)(e_H) = (h * y(j))(e_H)$ (since $y \in \mathrm{Map}_\psi \ldots$), and the last expression equals $y(j)(h)$, q.e.d.

$\alpha\beta$ is the identity: Let $z \in \mathrm{Map}_{\psi\pi}(J,S)$, $j \in J$. Then $(\alpha\beta(z))(j) = \beta(z)(j)(e_H) = z(\psi(e_H)j) = z(j)$, q.e.d. This completes the proof of (i).

(ii): We will give one argument for the general case, and another for the special case that H is abelian.

Note first that π^* commutes with faithfully flat base change, i.e. for any faithfully flat R-algebra T and any S in $\mathrm{GAL}(R,G)$, there is a canonical H-equivariant isomorphism $\pi^*(T \otimes S) \cong T \otimes \pi^* S$. By faithfully flat descent, it thus suffices to find such a T with $\pi^*(T \otimes S)$ an H-Galois extension of T. Taking $T = S$ and changing notation, we are reduced to proving: π^* of the trivial G-extension $R^{(G)}$ is an H-Galois extension of R. Let ι: $\{e\} \to G$, ι': $\{e\} \to H$ be the obvious maps. One checks quite easily: $\iota^* R$ is the trivial G-extension $R^{(G)}$. Since $\pi\iota = \iota'$, we obtain:

$$\pi^*(R^{(G)}) \cong \pi^*\iota^* R \cong (\iota')^* R \qquad (\text{by } (i))$$
$$\cong R^{(H)},$$

and we already know that this is indeed an H-Galois extension, q.e.d.

The following nice argument for H abelian is due to Harrison. We factor π as $\pi = \delta\gamma$, with $\gamma = (\mathrm{id}_G, e_H)$: $G \to G \times H$, and $\delta = (\pi, \mathrm{id}_H)$: $G \times H \to H$. Then γ is a split monomorphism, and δ is onto. It is sufficient to show (ii) for γ, and for δ, taking into account (i). For $\pi = \gamma$, one sees directly that $\pi^* S \cong S \otimes_R R^{(H)}$, with the obvious action of $G \times H$, and one can check that this is a $G \times H$-Galois extension. For $\pi = \delta$, i.e. π onto, one calculates from the definition that $\delta^* S = S^{\mathrm{Ker}(\delta)}$, which is indeed a Galois extension with group $\mathrm{Im}(\delta)$ by Thm. 2.2.

We now present Harrison's construction which makes the set $\mathrm{H}(R,G)$ into an abelian group if G is a finite *abelian* group. This will then be called the Harrison group of R and G. (Recall that $\mathrm{H}(R,G) = \mathrm{GAL}(R,G)/\cong$). We use without further comment the following easy fact: If S, $T \in \mathrm{GAL}(R,G)$, then $S \otimes_R T$ with the natural action of $G \times G$, is a $G \times G$-Galois extension of R. Let G be finite abelian, ι: $\{e\} \to G$ be the inclusion of the trivial group in G, μ: $G \times G \to G$ the multiplication (a homomorphism!), and j: $G \to G$ the map $g \longmapsto g^{-1}$ (again, a homomorphism).

Definition. The Harrison product $S \cdot T$ of S, $T \in \mathrm{GAL}(R,G)$ is defined to be

$$S \cdot T \;=\; \mu^*(S \otimes_R T) \in \mathrm{GAL}(R,G).$$

By functoriality, the Harrison product $[S \cdot T]$ of two isomorphism classes $[S]$, $[T]$ $\in \mathrm{H}(R,G)$ is a well-defined element of $\mathrm{H}(R,G)$. We shall often abuse notation and write $S \in \mathrm{H}(R,G)$ etc.

Theorem 3.2. [Harrison (1965)] *a) With this definition,* $\mathrm{H}(R,G)$ *becomes an abelian group whose neutral element is (the class of) the trivial extension* $R^{(G)}/R$.

b) If $\pi: G \to H$ *is a homomorphism from* G *to another abelian group* H*, then* $\pi^*: \mathrm{H}(R,G) \longrightarrow \mathrm{H}(R,H)$ *is a group homomorphism.*

Proof. a) This is a rather formal argument exploiting the functoriality properties. Let us begin by showing associativity of the Harrison product. Let $S, T, U \in \mathrm{H}(R,G)$.

Then:
$$\begin{aligned}(S\cdot T)\cdot U &= \mu^*\big(\mu^*(S\otimes_R T)\otimes_R U\big)\\ &= \mu^*\big((\mu^*\otimes \mathrm{id}_G{}^*)(S\otimes T\otimes U)\big)\\ &= (\mu(\mu\times\mathrm{id}))^*\,(S\otimes T\otimes U) \quad (3.1.\ \mathrm{b}))\\ &= (\mu(\mathrm{id}\times\mu))^*(S\otimes T\otimes U) \quad \text{(this is just the associativity of } G)\\ &= S\cdot(T\cdot U) \quad \text{(same calculation backwards).}\end{aligned}$$

In the same manner, one proves commutativity: $S\cdot T = \mu^*(S\otimes T) = (\mu\tau)^*(S\otimes T)$ (where $\tau: G\times G \to G\times G$ is the interchange isomorphism; $\mu = \mu\tau$ since G is commutative). Now $(\mu\tau)^*(S\otimes T) = \mu^*\tau^*(S\otimes T)$, and $\tau^*(S\otimes T) = T\otimes S$ by a direct argument. This finally gives $S\cdot T = T\cdot S$.

To see that $E = R^{(G)}$ is neutral, we recall that $E = \iota^*R$, where R is the (!) Galois extension of R with group $\{e\}$. For any $S \in \mathrm{H}(R,G)$, we then get:

$$\begin{aligned}S\cdot E = \mu^*(S\otimes_R \iota^*R) &= \mu^*(\mathrm{id}_G\times\iota)^*(S\otimes_R R)\\ &= \mathrm{id}_G{}^*(S\otimes_R R) \quad (\text{since } \mu(\mathrm{id}_G\times\iota) = \mathrm{id}_G)\\ &= S \quad \text{(direct argument).}\end{aligned}$$

Finally, we use j to construct an inverse of $S \in \mathrm{H}(R,G)$. We need a small auxiliary formula, to wit: $S\otimes S$ (as a $G\times G$–Galois extension) is isomorphic to Δ^*S, where $\Delta: G \to G\times G$ is the diagonal embedding. [Proof of the latter formula: There is an isomorphism $\alpha: \mathrm{Map}(G, S) \longrightarrow \mathrm{Map}_\Delta(G\times G, S)$ which maps $(f: G \to S)$ to $\big((g,h) \longmapsto g(f(g^{-1}h))\big)$. Furthermore, $\mathrm{Map}_\Delta(G\times G, S)$ is just Δ^*S. On the other hand, there is the canonical isomorphism $h: S\otimes S \longrightarrow \mathrm{Map}(G,S)$ from the definition of G–Galois extension. One then checks that the composite αh is $G\times G$–equivariant.] Now we claim that j^*S gives an inverse to S in the Harrison group. We calculate:

$$\begin{aligned}S\cdot\sigma^*S = \mu^*(S\otimes\sigma^*S) &= \mu^*((\mathrm{id}_G{}^*\otimes j^*)(S\otimes S))\\ &= \mu^*(\mathrm{id}_G\times j)^*\Delta^*(S)\\ &= (\mu(\mathrm{id}\times j)\Delta)^*(S)\\ &= (\iota\varepsilon)^*(S) \quad \text{with } \varepsilon: G \to \{e\} \text{ the obvious map}\\ &= \iota^*(\varepsilon^*(S))\\ &= \iota^*(S^G) = \iota^*R = R^{(G)} \quad \text{(cf. above), q.e.d.}\end{aligned}$$

The proof of b) is quite similar, and hence omitted.

By Thm. 3.1 and 3.2, we have, for any commutative ring, a functor H(R,–) from the category of finite abelian groups to the category of abelian groups. The most important general property of this functor is the following:

Theorem 3.3. *The functor* H(R,–) *is left exact, i.e.:* $0 \to J \overset{i}{\to} G \overset{\psi}{\to} G' \to 0$ *is a short exact sequence of finite abelian groups, then the induced sequence*

$$0 \longrightarrow \mathrm{H}(R,J) \overset{i^*}{\longrightarrow} \mathrm{H}(R,G) \overset{\psi^*}{\longrightarrow} \mathrm{H}(R,G')$$

is exact.

Remark. H(R,–) is usually not right exact; one can establish a relation between the obstruction to right-exactness and certain Brauer groups. We shall not pursue this.

Proof of 3.3. Since ψi is the zero map, $(\psi i)^*$ has also to be the zero map. Hence $\psi^* i^* = 0$. For the other parts of the proof, we need a lemma.

Lemma 3.4. $S \in \mathrm{GAL}(R,G)$ *is trivial iff there exists an* R*-algebra homomorphism* $S \overset{\varepsilon}{\to} R$. *(Such an* ε *is called an* augmentation.*)*

Proof of Lemma 3.4. The trivial extension $R^{(G)}$ obviously has augmentations, namely, the canonical projections to R. Suppose on the other hand that ε: $S \to R$ is an augmentation. Define an R-algebra homomorphism(!) φ: $S \to R^{(G)}$ by $\varphi(y) = (\varepsilon(\sigma y))_{\sigma\in G}$. It is easy to check that φ is also G-equivariant, hence an isomorphism by 1.12.

We continue in the proof of Thm. 3.3. Suppose $S \in \mathrm{H}(R,G)$ with i^*S trivial. Then, by the lemma, i^*S has an augmentation ε into R. Recall that $i^*S = \mathrm{Map}_i(G,S)$. We can construct an R-algebra homomorphism α: $S \to i^*S$ by letting , for $y \in S$, and $\sigma \in G$: $\alpha(y)(\sigma) = \sigma(y)$ if $\sigma \in \mathrm{Im}(i)$, and $\alpha(y) = 0$ otherwise. Hence S itself has an augmentation $\varepsilon\alpha$, and hence is trivial, by the lemma.

To conclude the proof, we suppose $T \in \mathrm{H}(R,G)$ such that ψ^*T ($= T^{\mathrm{Ker}(\psi)}$) is trivial, i.e. $\psi^*T = R^{(G')}$. Recall $R^{(G')} = \bigoplus_{\sigma'\in G'} f_{\sigma'}\cdot R$, where $\{f_{\sigma'}\}$ is the canonical basis made up of idempotents. Identify $T^{\mathrm{ker}(\psi)}$ with $R^{(G')}$ and consider the R-algebra $A = T\cdot f_{e'}$ (where e' is the neutral element of G'). Now one checks, by reduction to the case $T = R^{(G)}$ (faithfully flat descent), that A is canonically a J-Galois extension of R. (Note $J = \mathrm{Ker}(\psi)$.) We define an R-algebra homomorphism β: $T \to i^*A$ by

$$\beta(y)(\sigma) = \sigma(y)\cdot f_{e'} \quad (y \in T,\ \sigma \in G).$$

To see that this is well-defined, one has to verify that $\beta(y)$ is indeed in $\mathrm{Map}_i(G,A)$. For $\tau \in J$ we have $\beta(y)(i(\tau)\cdot\sigma) = (i(\tau)\sigma)(y)\cdot f_{e'} = \tau(\sigma(y)\cdot f_{e'})$ (since $f_{e'}$ is fixed by τ), and the last expression is $\tau(\beta(y)(\sigma))$. Furthermore, β is G-equivariant. Again by 1.12, β is an isomorphism, and T is in the image of i^*, q.e.d.

This theorem has an important consequence:

Theorem 3.5. *Given a commutative ring R, there exists a profinite abelian group Ω_R such that the functor* $H(R,-)$: *{finite abelian groups}* $\longrightarrow$ *{abelian groups} is pro-represented by Ω_R, i.e. naturally isomorphic to* $\mathrm{Hom}_{cont}(\Omega_R,-)$.

Proof. Every left exact functor from finite abelian groups to abelian groups is pro-representable. For details, see Harrison (1965), proof of Thm. 4.

Remarks. a) The proof gives at the same time the *uniqueness* of Ω_R. Often Ω_R is called the *abelian fundamental group* of R, or the *abelianized absolute Galois group* of R. Explicitly one has $\Omega_R = \mathrm{proj.lim}\big(\,H(R,\mathbb{Z}/n\mathbb{Z})^{\vee}\big)$, where the groups $\mathbb{Z}/n\mathbb{Z}$ form an inductive system indexed by the divisibility lattice of $\mathbb{N}$, and $^{\vee}$ means Pontryagin dual. See Harrison, loc.cit.

b) By the methods of the proof of 3.2, one can see without difficulty: if $[k]$: $G \to G$ is the homomorphism $g \longmapsto g^k$ (G finite abelian), then $[k]^*$: $H(R,G) \to H(R,G)$ is multiplication by k in the Harrison group. In particular, $H(R,G)$ is annihilated by the exponent of G, hence torsion. Therefore the Pontryagin dual in a) is indeed profinite.

We want to show that the group Ω_R has in many cases an interpretation as a (profinite) group of automorphisms of an appropriate (infinite) extension of R. If $R = K$ is a field, the required group is $\mathrm{Aut}(K^{ab}/K)$, where K^{ab} is the maximal abelian Galois extension of K. More generally, Janusz (1966) has proved the existence of a *separable closure* for every connected ring R. This is by definition a connected R-algebra R^{sep} which is a filtered union of G-Galois extensions of R (G varies, of course), such that every connected Galois extension S/R is embeddable in R^{sep}. There is no ambiguity here as to what the Galois group of S/R is, because of the following result (see Chase-Harrison-Rosenberg (1965), Cor. 3.3 or this chapter, 7.3): If S/R is a G-Galois extension of connected rings, then $\mathrm{Aut}(S/R) = G$. As proved by Janusz (1966), the group $\Psi_R = \mathrm{Aut}(R^{sep}/R)$ is a filtered projective limit of finite groups (more precisely: of Galois groups of Galois extensions contained in R^{sep}), hence Ψ_R is profinite. (For another exposition of this material, see also DeMeyer-Ingraham (1971).)

Our next objective is the following result.

Theorem 3.6. *Let R be connected, and $\Psi_R = \mathrm{Aut}(R^{sep}/R)$ as in the preceding discussion. Then there is a natural isomorphism*

$$\nu\colon \mathrm{Hom}_{cont}(\Psi_R,-) \longrightarrow H(R,-)$$

of functors from finite abelian groups to abelian groups.

We need a few preparations for the proof.

Lemma 3.7. *Let R be connected, S/R, T/R be two Galois extensions (with maybe different groups), and f: $S \longrightarrow T$ an R-algebra homomorphism. Then* Ker(f) *is generated by an idempotent.*

Proof. It is well known that an ideal I of S is generated by an idempotent iff I is finitely generated and $I^2 = I$. [One direction is clear. If I is finitely generated and $I \cdot I = I$, then by Bourbaki, Alg. comm., II §2 n^0 2 cor. 3, there exists $e \in I$ such that $(1-e)I = 0$. It follows easily that $e^2 = e$ and $I = eS$.] Both of these properties of I may be tested by a faithfully flat base extensions, hence we may, and shall, assume that both S and T are trivial, i.e. a finite product of copies of R, considered just as R-algebras. Here the desired property of Ker(f) is easily obtained by explicit calculations with canonical idempotents.

Remark. In the proof, we only used that S and T become isomorphic to a product of copies of the ground ring after suitable faithfully flat base change.

Proposition 3.8. [Harrison (1965)] *Let R be connected, S/R a G-Galois extension with finite abelian group G. Then there exists a subgroup $H \subset G$ and a* <u>*connected*</u> *H-Galois extension U/R such that $S \cong i^*U$ (where i: $H \to G$ is the inclusion).*

Proof. Since R is connected, every finitely generated projective R-module has a well-defined rank ($\in \mathbb{N}$). From this it follows that the R-algebra S can be decomposed as a product of finitely (at most rank(S)) algebras without proper idempotents, i.e. there are irreducible orthogonal idempotents $e_1,\dots,e_n$ in S whose sum is 1, the set M of irreducible idempotents equals $\{e_1,\dots,e_n\}$, and $S = \bigoplus_{i=1}^n e_iS$, where all e_iS are connected.

G operates on M. Since the sum over any G-orbit on M is G-invariant, hence an idempotent of R, the operation of G on M must be transitive. Pick $e_1 \in M$, and let H be the stabilizer of e_1 in G. Since G is abelian, H is the stabilizer of every $e \in M$. Let $U = e_1S$. By transitivity of G on M, all eS ($e \in M$) are isomorphic to U as R-algebras, so $S \cong U^{(M)}$ (noncanonically), U is connected, and H acts on U by R-automorphisms. We now claim that U/R is H-Galois.

First of all, H embeds into Aut(U). (If $\sigma \in H$ is identity on $U = e_1S$, one sees, again since G is transitive on M, that σ is identity on all eS ($e \in M$), hence $\sigma =$ id.) Next, G/H operates transitively and without fixed points on M, whence $[G{:}H] = |M|$. From the definition of Galois extension, one has $\mathrm{rank}_R(S \otimes S) = \mathrm{rank}_R(S)\cdot|G|$, hence $\mathrm{rank}_R(S) = |G|$. Therefore $\mathrm{rank}_R(U) = |H|$ (use $S \cong U^{(M)}$). Thus we know that $U[H]$ and $\mathrm{End}_R(U)$ are finitely generated projective of the same rank over R, and it suffices to show that the map j: $U[H] \longrightarrow \mathrm{End}_R(U)$ (cf. Thm. 1.6) is *onto*. U is embedded in $\mathrm{End}_R(U)$ via left multiplications. With this identification, one has the following formulas in $\mathrm{End}_R(U)$: $e_1{\cdot}\sigma{\cdot}e_1 = e_1{\cdot}\sigma = \sigma{\cdot}e_1$ for $\sigma \in H$, $e_1\sigma e_1 = 0$ for σ not in H. Let $\varphi \in \mathrm{End}_R(U)$. Then $e_1{\cdot}\varphi{\cdot}e_1 \in \mathrm{End}_R(S)$; by 1.6 (*iii*), there are $s_\sigma \in S$

with $\Sigma_{\sigma\in G} s_\sigma\cdot\sigma = e_1\cdot\varphi\cdot e_1$. Multiplying this equation with e_1 from both sides, we obtain by the above remarks: $\Sigma_{\sigma\in H}(e_1 s_\sigma)\cdot\sigma\cdot e_1 = e_1\cdot\varphi\cdot e_1$. If we restrict both sides to U, we get the equation $\Sigma_{\sigma\in H}(e_1 s_\sigma)\cdot\sigma = \varphi$ in $\mathrm{End}_R(U)$, hence j is onto.

It remains to show $S \cong i^*U$. This is done by the usual method: Define a map $\beta\colon S \longrightarrow i^*U$ by $\beta(s)(\sigma) = e_1\sigma(s)$. Exactly as in the proof of 3.3 one sees that β is a G-equivariant R-algebra homomorphism, hence an isomorphism by 1.12. Q.E.D.

Remark. If $R = K$ is a field, then one can show that the connected H-extension U is also a field. U is called the "Kernkörper" (core field) of S/K in Hasse (1949).

Proof of 3.6. We define for each finite abelian group G a map $\nu_G : \mathrm{Hom}_{cont}(\Psi_R, G) \longrightarrow \mathrm{H}(R,G)$. Recall R^{sep} is a separable closure of R. For any continuous homomorphism $f\colon \Psi_R = \mathrm{Aut}(R^{sep}/R) \longrightarrow G$, we define

$$\nu_G(f) = f_E{}^*(E) \in \mathrm{H}(R,G),$$

where E is any abelian Galois extension of R, contained in R^{sep}, such that f factors through a homomorphism $f_E\colon \mathrm{Aut}(E/R) \longrightarrow G$. (The existence of such an E follows, since f is continuous, and G is abelian.) Of course we must show that this is independent of the choice of E: if E' is another, w. l. o. g. $E \subset E' \subset R^{sep}$, then by the Main Theorem (§2), the Galois group of E' maps onto the Galois group of E (call this epimorphism π), and E is the fixed field of $\mathrm{Ker}(\pi)$ in E'. From $f_{E'} = f_E\pi$ and $E = \pi^*E'$ we get $f_{E'}{}^*E' = f_E{}^*E$.

Now we prove that ν_G preserves sums. Let us write G additively for this. Let $f, g \in \mathrm{Hom}_{cont}(\Psi_R,G)$, and let $\cdot$ denote the Harrison product. We obtain:

$$\begin{aligned}
\nu_G(f)\cdot\nu_G(g) &= f_E{}^*E\cdot g_E{}^*E \qquad (E\subset R^{sep} \text{ Galois}/R, \text{ large enough})\\
&= \mu^*(f_E{}^*E \otimes_R g_E{}^*E) \qquad (\text{def. of } \cdot\,)\\
&\cong \mu^*\big((f_E\times g_E)^*(E\otimes_R E)\big)\\
&\cong (\mu^*(f_E\times g_E)^*\Delta^*)(E) \qquad (\Delta^*E \cong E\otimes E, \text{ cf. proof of 3.2})\\
&= (f_E+g_E)^*E,
\end{aligned}$$

the last step being justified because $\mu(f_E\times g_E)\Delta$ coincides with f_E+g_E.

The proof of the naturality of ν_G in G is similar (and easier), and we omit it.

We show ν_G is injective: Suppose f^*E is the trivial G-extension (where f is in $\mathrm{Hom}(\mathrm{Aut}(E/R),G)$, E/R abelian connected Galois extension). We can factor f in the form $i\psi$, with ψ surjective and i injective. By 3.3, already ψ^*E is trivial, i.e. $E^{\mathrm{Ker}(\psi)}$ is a product of copies of R, as an R-algebra, but also connected (since E is). Therefore we must have $E^{\mathrm{Ker}(\psi)} = R$, and for reasons of rank: $\mathrm{Coker}(\psi)$ is trivial. Hence f is the zero homomorphism.

ν_G is surjective: Suppose we are given a G-Galois extension S/R. If S is connected, then we may assume $S \subset R^{sep}$, and there is a natural continuous epimorphism $f\colon \Psi_R \longrightarrow \mathrm{Aut}(S/R) = G$. Taking $E = S$, we get the identity for f_E, hence $\nu_G(f) = S$. This was the easy case. The problem arises when S is not connected. But then, thanks to 3.8, we find an inclusion $i\colon H \subset G$ of a subgroup H and a connected H-Galois extension U/R with $i^*U \cong S$. By naturality of ν, it then suffices to find a preimage of U under ν_H, but this we can do since U is connected. Q.E.D.

Corollary 3.9. *For any connected (commutative) ring R, the abelian fundamental group Ω_R is isomorphic to the abelianization $\Psi_R/[\Psi_R,\Psi_R]$.*

Proof. By Thm. 3.6, the profinite group $\Psi_R/[\Psi_R,\Psi_R]$ pro-represents the functor $\mathrm{H}(R,-)$. The group Ω_R pro-represents this functor by definition. It is well-known from category theory that a pro-representing object of a given functor is unique up to isomorphism.

At the end of this section, we give some results which show how the functorialities of the two arguments in $\mathrm{H}(R,G)$ interact.

Proposition 3.10. *Let $f\colon R \to T$ be a homomorphism of (commutative) rings, and $\pi\colon G \to H$ a homomorphism of finite groups. Then the diagram*

$$\begin{array}{ccc} \mathrm{H}(R,G) & \xrightarrow{\pi^*} & \mathrm{H}(R,H) \\ {\scriptstyle T\otimes -}\big\downarrow & & \big\downarrow{\scriptstyle T\otimes -} \\ \mathrm{H}(T,G) & \xrightarrow{\pi^*} & \mathrm{H}(T,H) \end{array}$$

is commutative.

Proof. If S is flat over R, the proposition follows directly from the definitions. The reader will probably note that we used the result (for S/R faithfully flat) in the proof of Thm. 3.1 already.

For the general case, one exhibits for $S \in \mathrm{H}(R,G)$ a canonical map

$$\alpha_{S/R}\colon\ T\otimes_R \pi^*S \longrightarrow \pi^*(T\otimes_R S);$$

one checks that the definition of α is compatible with faithfully flat base change, which reduces the proof to the case where S is a trivial G-extension. This case can easily be done directly.

Corollary 3.11. *If R, S, G are as in 3.10, and G is abelian, then the map $T\otimes_R -\colon \mathrm{H}(R,G) \longrightarrow \mathrm{H}(S,G)$ is a group homomorphism.*

Proof. The Harrison product was defined with the help of μ^* ($\mu\colon G\times G \to G$ the multiplication map.) By 3.10, the map $T\otimes_R -$ commutes with μ^*. From this the claim follows easily.

Theorem 3.12. *Let T/R be an H–Galois extension of connected rings, H abelian. Then for every finite abelian group, the kernel of $T\otimes_R-$: $\mathrm{H}(R,G) \longrightarrow \mathrm{H}(T,G)$ is canonically isomorphic to* $\mathrm{H}(R,H)$.

Proof. Define α: $\mathrm{Hom}(H,G) \longrightarrow \mathrm{H}(R,G)$ by $\alpha(\pi) = \pi^*T$ for π: $H \to G$. One checks that α is a group homomorphism. If $\alpha(\pi)$ is trivial, then by 3.3 already ψ^*T is trivial, where ψ: $H \to \mathrm{Im}(\pi)$ is the surjection defined by π. But ψ^*T is (up to canonical isomorphism) a subalgebra of T, and T is connected. Hence necessarily $\psi^*T = R$, and π is the trivial homomorphism.

For each $\pi \in \mathrm{Hom}(H,G)$ we have $T\otimes_R \pi^*T \cong \pi^*(T\otimes_R T)$ by 3.8, and already $T\otimes T/T$ is the trivial H–extension. Hence the composite map $(T\otimes-)\alpha$ is zero.

To conclude the proof, assume $S \in \mathrm{H}(R,G)$ such that $T\otimes_R S$ is the trivial G–extension. By 3.8, we can find an inclusion i: $G' \subset G$ of a subgroup and a connected G'–Galois extension S' such that $S \cong i^*S'$. Then $T\otimes_R S \cong i^*(T\otimes_R S')$, and by 3.3, $T\otimes_R S'/T$ is the trivial G'–extension. This implies (by 3.4) that there exists a T–algebra homomorphism $T\otimes_R S' \to T$, hence an R–algebra homomorphism φ: $S' \to T$. Since S' is connected, φ must be injective by 3.7. Since S' is separable over R (every Galois extension is separable!), we may use the main theorem 2.3 and obtain: $S' \cong \psi^*T$ for some surjection $H \to G'$. Putting things together, we obtain $S = i^*S' = i^*\psi^*T = \pi^*T$ for $\pi = i\psi$: $H \to G$.

§4 Ramification

We have not seen many examples of Galois extensions yet. The purpose of this section is to set up a translation machinery which allows us to find many Galois extensions of rings of a number–theoretical kind. These results are quite well known to experts, but it seems convenient to give proofs here.

Notation: If K is an algebraic number field (i.e. a finite extension of $\mathbb{Q}$), then $\mathcal{O}_K$ denotes the ring of integers in K. Example: $\mathcal{O}_{\mathbb{Q}} = \mathbb{Z}$.

Remark. If L/K is a G–Galois extension of number fields, then G operates also on the ring $\mathcal{O}_L$, and $(\mathcal{O}_L)^G = \mathcal{O}_L \cap K = \mathcal{O}_K$.

The remark suggests a possibility of $\mathcal{O}_L/\mathcal{O}_K$ being a G–Galois extension of rings. Some care is necessary, as is shown by the following simple example:

Example. *Let $K = \mathbb{Q}$, $L = \mathbb{Q}(i)$ (with $i^2 = -1$). Then $G = \{\mathrm{id}, \sigma\}$, and $\mathcal{O}_L/\mathcal{O}_K$ is* <u>*not*</u> *a G-Galois extension (with the canonical action of G, of course).*

Proof. We have $\mathcal{O}_L = \mathbb{Z}[i]$. We shall show that the map j (see 1.3, 1.6) from $\mathbb{Z}[i]*G$ to $\mathrm{End}_{\mathbb{Z}}(\mathbb{Z}[i])$ is not surjective. (It is injective, by the way.) Let $\mathfrak{p}$ be the maximal ideal $\mathfrak{p} = (1+i)$ of $\mathbb{Z}[i]$. Note $2 \in \mathfrak{p}$. Claim: For each $\varphi \in \mathrm{Im}(j)$, $\varphi(1) \equiv \varphi(i) \pmod{\mathfrak{p}}$. (If we have this, then we also know that j is not onto, because there certainly exists $\varphi \in \mathrm{End}(\mathbb{Z}[i])$ with $\varphi(1) = 1$, $\varphi(i) = 0$.) Proof of claim: Write $\varphi = j(a\cdot\mathrm{id} + b\cdot\sigma)$, $a, b \in \mathbb{Z}[i]$. We get: $\varphi(1) = a+b$, $\varphi(i) = a\cdot i - b\cdot i \equiv a-b \pmod{\mathfrak{p}}$, and $a-b \equiv a+b$, mod 2, and a fortiori mod $\mathfrak{p}$.

As some readers may already have guessed, it is ramification which prevents $\mathcal{O}_L$ from being a Galois extension in the above example. (We use without further explanation the basic notions of ramification, trace, norm, and discriminant in algebraic number theory.) To state the relevant theorem, we need the notion of S-integers. Let S be any set of finite places of the number field K (equivalently: S is a set of nonzero prime ideals of $\mathcal{O}_K$). Then the ring of *S-integers* $\mathcal{O}_{K,S}$ of K is defined as follows:

$$\mathcal{O}_{K,S} = \{x \in K \mid v_{\mathfrak{p}}(x) \geq 0 \text{ for all } \mathfrak{p} \text{ not in } S\}.$$

We allow $v_{\mathfrak{p}}(x) = \infty$, so 0 is in $\mathcal{O}_{K,S}$. For empty S, $\mathcal{O}_{K,S}$ is just $\mathcal{O}_K$.

For a G-Galois extension L/K and a set S of finite places of K, we set S' equal to the set of all places $\mathfrak{q}$ of L that divide some $\mathfrak{p} \in S$. Again, G acts on $\mathcal{O}_{L,S'}$, and the ring of G-invariants $\mathcal{O}_{L,S'}$ coincides with $\mathcal{O}_{K,S}$.

Theorem 4.1. [cf. Auslander - Buchsbaum (1959)] *Let L/K be a G-Galois extension of algebraic number fields, S and S' as above. Then $\mathcal{O}_{L,S'}/\mathcal{O}_{K,S}$ is a G-Galois extension if and only if L/K is unramified at all finite places which are not in S.*

Proof. We first perform a reduction to the local case. For this, assume we are given an R-algebra T which is finitely presented as an R-module, and on which a group G acts faithfully by R-algebra automorphisms. It is then a consequence of standard localization techniques (and the definition 1.5, of course) that T/R is G-Galois iff $T_{\mathfrak{M}}/R_{\mathfrak{M}}$ is G-Galois for all maximal ideals $\mathfrak{M}$ of R. This is applicable in our situation with $R = \mathcal{O}_{K,S}$ and $T = \mathcal{O}_{L,S'}$. The maximal ideals of R correspond to the maximal ideals $\mathfrak{p}$ of $\mathcal{O}_K$ outside S, and the localization of R with respect to such a maximal ideal is just $(\mathcal{O}_K)_{\mathfrak{p}}$. It therefore suffices to prove for an arbitrary maximal ideal $\mathfrak{p}$ of $\mathcal{O}_K$:

$$(*) \qquad (\mathcal{O}_L)_{\mathfrak{p}}/(\mathcal{O}_K)_{\mathfrak{p}} \text{ is } G\text{-Galois} \iff \mathfrak{p} \text{ does not ramify in } L/K.$$

Let us change notation as follows: $R = (\mathcal{O}_K)_{\mathfrak{p}}$, $T = (\mathcal{O}_L)_{\mathfrak{p}}$. We consider the map h of Def. 1.5. Since R is local, we may pick an R-basis $x_1,\dots,x_n$ of T. (Then $n = \mathrm{rank}_R(T) = [L:K] = |G|$.) As we saw in the proof of 1.6, the representing matrix of

h: $T \otimes_R T \longrightarrow T^{(G)}$ is $A = (\sigma(x_i))_{\sigma,i}$, where σ runs over G, and i from 1 to n. We thus have to decide whether A is invertible in the $n \times n$ matrix ring over T. One calculates that $A^t \cdot A = (tr(x_i x_j))_{i,j}$, with tr: $L \to K$ the usual trace ($tr(y) = \sum_{\sigma \in G} \sigma(y)$). By construction of the local discriminant $\mathrm{disc}_{\mathfrak{p}}(L/K)$, this discriminant is precisely generated by $\det(A^t \cdot A)$ as an ideal of R. It is also well known that this discriminant is the unit ideal if and only if L/K is unramified at $\mathfrak{p}$. Hence: L/K is unramified at $\mathfrak{p}$ $\Longleftrightarrow$ $A^t \cdot A$ is invertible $\Longleftrightarrow$ A is invertible $\Longleftrightarrow$ h is an isomorphism $\Longleftrightarrow$ T/R is G-Galois.

Corollary 4.2. *For abelian G, the canonical map α: $\mathrm{H}(R,G) \to \mathrm{H}(K,G)$ is injective, and its image consists of the G-extensions L/K, whose core field (= connected part) is unramified over K outside S and infinity. For the injectivity of α, it suffices that R is an integrally closed domain with field of quotients K.*

Proof. For any $A \in \mathrm{H}(R,G)$, A coincides with the integral closure of R in $K \otimes_R A$ by Harrison (1965) [Thm. 5]. (One uses the trace tr: $K \otimes_R A \longrightarrow K$, and the fact that the trace of an R-integral element is again R-integral, hence in R.) This gives at once the injectivity of α. If $L \in \mathrm{H}(K,G)$ is a field, then the only possible preimage of L under α is the integral closure of R in L, i.e. the ring of S'-integers of L. Hence $L \in \mathrm{Im}(\alpha)$ iff L is unramified outside S and infinity. The case L not a field is treated by means of Prop. 3.8.

§5 Kummer theory and Artin-Schreier theory

The theories mentioned in this heading are first steps towards a classification of *all* G-Galois extensions of a given commutative ring R with given finite abelian group G. We begin with Kummer theory. This is quite classical and well-known for fields. The "ring case" can be found in several sources (Borevich (1979), Waterhouse (1987), Milne (1986)). Nevertheless we found it worthwile to include down-to-earth proofs (not using cohomology) which are at the same time reasonably short. We need a definition which looks technical at first glance:

Definition. Let $n \in \mathbb{N}$. A commutative ring R is *n-kummerian*, if it contains n^{-1} and a root ζ of the n-th cyclotomic polynomial $\Phi_n \in \mathbb{Z}[X]$. (There may then be several roots of Φ_n in R, but usually one of them will be fixed once for all.)

One then may construct Galois extensions with Galois group C_n = (cyclic group of order n with generator σ) as follows: Let R be n-kummerian, and suppose u is a unit of R. Define

$$R(n, u) \;=\; R[X]/(X^n - u), \text{ with } C_n\text{-action given by } \sigma\overline{X} = \zeta\cdot\overline{X}.$$

The action makes sense, since the R-algebra endomorphism σ': $X \longmapsto \zeta X$ of $R[X]$ maps $X^n - u$ to $\zeta^n X^n - u = X^n - u$, and since the n-th power of σ' is the identity. Sometimes when it seems safe, we shall write $\sqrt[n]{u}$ or $u^{1/n}$ for $\overline{X}$.

Lemma 5.1. *a) $R(n; u)$ is a C_n-Galois extension of R.*

b) If $v \in R^*$ is another unit, then the Harrison product $R(n; u)\cdot R(n; v)$ is isomorphic to $R(n; uv)$.

*c) The map $u \longmapsto R(n; u)$ induces a monomorphism $i = i_R$: $R^*/p^n \longrightarrow \mathrm{H}(R, C_n)$. (**Notation**: For M any abelian group, we let M/p^n denote M/p^nM, or M/M^{p^n}, according to whether M is an additive or multiplicative abelian group.)*

Proof. a) Let $\alpha = \overline{X}$. Then $S = R(n; u) = R \oplus \alpha R \oplus \ldots \oplus \alpha^{n-1}R$. For $1 \le i \le n-1$, $\zeta^i - 1$ is a unit in R. (Reason: If ζ_n denotes a primitive nth root of unity in $\mathbb{C}$, then $(1-\zeta_n)(1-\zeta_n^2)\cdot\ldots\cdot(1-\zeta_n^{n-1}) = \big((X^n-1)/(X-1)\big)\big|_{X=1} = n$, hence $(1-\zeta_n^i)$ is a unit in $\mathbb{Z}[\zeta_n, n^{-1}]$, and there is a ring homomorphism from $\mathbb{Z}[\zeta_n, n^{-1}]$ to R which maps ζ_n to ζ.) From this one sees that the fixed ring of σ in S is R, since σ operates on the cyclic summand $\alpha^i R$ as multiplication by ζ^i.

One now calculates the representing matrix A of the map h: $S \otimes S \longrightarrow S^{(C_n)}$ with respect to the basis $1, \alpha, \ldots, \alpha^{n-1}$ and obtains

$$A \;=\; \begin{pmatrix} 1 & \alpha & \ldots & \alpha^{n-1} \\ 1 & \zeta\alpha & \ldots & \zeta^{n-1}\alpha^{n-1} \\ \cdot & \cdot & \ldots & \cdot \\ \cdot & \cdot & \ldots & \cdot \\ 1 & \zeta^{n-1}\alpha & \ldots & \zeta^{(n-1)^2}\alpha^{n-1} \end{pmatrix}.$$

Hence $\det(A) = \alpha^{n(n-1)/2}$ times the Vandermonde determinant $\det\big(\zeta^{ij}\big)_{0\le i,j<n}$. The first factor is a unit in S because α is (recall $\alpha^n = u$). The second factor is also a unit in R, since again all $\zeta^i - \zeta^j$ $(0 \le i < j < n)$ are units in R. Hence by definition, S/R is C_n-Galois. (The fact that C_n operates faithfully was proved along the road: $\zeta^i \neq \zeta^j$ for $0 \le i < j < n$ suffices for this.)

b) This is a brute force calculation which uses the basis $(\alpha^i)_{0\le i<n}$ of $S = R(n; u)$ over R with $\alpha^n = u$ constructed in a), and the analogous basis (β^i) with $\beta^n = v$ in $T = R(n; v)$. Recall for this purpose that $S\cdot T = \mu^*(S \otimes_R T)$, where μ: $C_n \times C_n \to C_n$ is multiplication. Hence $S\cdot T$ is the fixed ring of $\mathrm{Ker}(\mu)$ in $S \otimes_R T$, and $\mathrm{Ker}(\mu)$ is generated by the pair $(\sigma, \sigma^{-1}) \in C_n \times C_n$. From this one calculates directly that $S\cdot T$ is the R-span of $\{1\otimes 1, \alpha\otimes\beta, \ldots, \alpha^{n-1}\otimes\beta^{n-1}\}$. Obviously the n-th power of $\alpha\otimes\beta$ equals uv, and from this the required isomorphism is easily obtained.

c) We have to show: $R(n;u)$ is trivial iff u is a n-th power in R^*. Now obviously the R-algebra homomorphisms ε: $R(n;u) \longrightarrow R$ are in bijection with the set of n-th roots of u in R (or R^*). Hence, this set is nonempty iff there exists one such ε, and this is equivalent to the triviality of $R(n;u)$ by Lemma 3.4. Q.E.D.

The question arises: what is the cokernel of i_R? In the process of answering this question, we also present a description of $\mathrm{H}(R,C_n)$ (for R n-kummerian) which is slightly more instructive than just the short exact sequence $1 \to \mathrm{Ker}(i_R) \to \mathrm{H}(R,C_n) \to \mathrm{Coker}(i_R) \to 0$.

Definition. Let R be a commutative ring, and recall that an *invertible* R-module is a finitely generated projective R-module of constant rank 1. A *discriminant module* (*of type n*) over R is a pair (M,φ) where M is an invertible R-module, and φ is an isomorphism $M^{\otimes n} \to R$. (The notation $M^{\otimes n}$ is shorthand for $M\otimes_R \dots \otimes_R M$, n factors.) Two discriminant modules (M,φ) and (M',φ') are *isomorphic*, if there exists an R-isomorphism f: $M \to M'$ such that $\varphi'(f^{\otimes n}) = \varphi$: $M^{\otimes n} \to R$.

Remark. The appellation "discriminant module" is unfortunate, but current in the literature, and I don't have any better suggestion.

Lemma 5.2. *The isomorphism classes of discriminant modules (of type n over R) form an abelian group* $\mathrm{Disc}(R,n)$ *in a natural way.*

Proof. The multiplication of discriminant modules is given by $(M,\varphi)\cdot(M',\varphi') = (M\otimes_R M', \varphi\otimes\varphi')$. In order that the map $\varphi\otimes\varphi'$ make sense, it is necessary to identify $(M\otimes_R M')^{\otimes n}$ with $M^{\otimes n}\otimes_R M'^{\otimes n}$, and $R\otimes_R R$ with R. Obviously, the class of (R,id_R) gives a neutral element, and associativity up to isomorphism is clear. If (M,φ) is given, we may construct an inverse by considering (M^*,φ^{*-1}) (here $(-)^*$ denotes R-duals). One has $(M,\varphi)\cdot(M^*,\varphi^{*-1}) \cong (R,\psi)$ for some ψ: $R \to R$, because $M\otimes_R M^* \cong R$. Now one can check that $\psi = \mathrm{id}_R$; alternatively, one shows that in any case, (R,ψ^{-1}) is inverse to (R,ψ) up to isomorphism. Note in this context that isomorphisms ψ: $R \to R$ are just multiplication by units of R.

Proposition 5.3. *For n-kummerian rings R, there is a canonical isomorphism*

$$d:\ \mathrm{H}(R,C_n) \longrightarrow \mathrm{Disc}(R,n).$$

Proof. Let χ be the character $C_n \to \langle\zeta\rangle$ given by $\sigma \longmapsto \zeta$. (Since ζ can be obtained as the image of ζ_n (a primitive nth root of unity in $\mathbb{C}$) under a ring homomorphism, we have $\zeta^n = 1$. Moreover all ζ^i, $i = 0,\dots,n-1$, are distinct, as shown in the proof of 5.1 a), so the exact multiplicative order of ζ is n.) For any module M over the group ring $R[C_n]$, we define for $0 \le i < n$:

$$\begin{aligned} M^{(i)} &= \{x \in M \mid \sigma x = \zeta^i\cdot x\} \\ &= \{x \in M \mid C_n \text{ operates on } x \text{ via the character } \chi\}. \end{aligned}$$

Since $n^{-1} \in R$, the quantities $e_i = n^{-1} \cdot \sum_{j=0}^{n} \chi^i(\sigma^{-j}) \cdot \sigma^j \in R[C_n]$ are well-defined. One checks that the e_i, $i = 0,\ldots,n-1$, are pairwise orthogonal idempotents whose sum is 1. (What one needs for this, is the formula $1+\zeta^i+\zeta^{2i}+\ldots+\zeta^{(n-1)i} = 0$ for $1 \le i \le n-1$, and this formula holds because it holds for ζ_n in the place of ζ, see beginning of this proof.) We have $\sigma \cdot e_i = \chi^i(\sigma) \cdot e_i = \zeta^i \cdot e_i$, hence M is the direct sum of all $e_i M$, and $e_i M = M^{(i)}$.

After these preparations, let $S \in \mathrm{H}(R,C_n)$ and consider the decomposition $S = \bigoplus_{i=0}^{n} S^{(i)}$. Note that $S^{(0)} = e_0 S = n^{-1} \cdot tr(S) = tr(S) = R$ by 1.10. We make two assertions:

a) All $S^{(i)}$ are invertible R-modules.

b) The multiplication in S induces isomorphisms $S^{(i)} \otimes_R S^{(j)} \longrightarrow S^{(i+j)}$ for all $i, j \in \mathbb{N}$. (The upper indices have to be read modulo n.)

These two assertions are proved by faithfully flat descent. We know beforehand that all $S^{(i)}$ are finitely generated projective over R, since they are direct summands of S, which is f.g. projective by 1.6. We also can see directly that $S^{(i)} S^{(j)}$ is automatically contained in $S^{(i+j)}$: take $x \in S^{(i)}$, $y \in S^{(j)}$. Then $\sigma(xy) = \sigma(x)\sigma(y) = \zeta^i \cdot x \cdot \zeta^j \cdot y = \zeta^{i+j} \cdot xy$, hence $xy \in S^{(i+j)}$. Thus two things are left to prove: $\mathrm{rank}(S^{(i)}) = 1$, and the maps $\mathrm{mult}_{i,j}$: $S^{(i)} \otimes_R S^{(j)} \longrightarrow S^{(i+j)}$ are isomorphisms. Let T be a faithfully flat R-algebra such that S_T $(= T \otimes_R S)$ is the trivial G-extension. ($S = T$ is a possibility.) It suffices to show that $T \otimes S^{(i)}$ has T-rank 1, and $T \otimes \mathrm{mult}_{i,j}$ is an isomorphism. But the definition of $S^{(i)}$ and $\mathrm{mult}_{i,j}$ is compatible with base change, hence we may simply assume S/R is the trivial G-extension: $S = R^{(G)}$. In this situation, we can calculate, and obtain

$$S^{(i)} = R \cdot f_i, \text{ with } \left(f_i \colon \sigma^k \longmapsto \zeta^{ik}\right) \in R^{(G)}.$$

From this description one sees immediately that $\mathrm{rank}_R(S^{(i)}) = 1$, and $\mathrm{mult}_{i,j}$ is an isomorphism because $f_i \cdot f_j = f_{i+j}$.

Let $P = S^{(1)}$. We obtain by iteration an isomorphism μ: $P^{\otimes n} \to S^{(n)} = S^{(0)} = R$. We can now define the map d: $\mathrm{GAL}(R,C_n) \longrightarrow \mathrm{Disc}(R,n)$ by $d(S) = (P,\mu)$. If α: $S \to S'$ is an isomorphism of Galois extensions, then f induces an isomorphism $P = S^{(1)} \to S'^{(1)} = P'$, and also an isomorphism of discriminant modules $(P,\mu) \to (P',\mu')$. Hence we obtain a well-defined map d: $\mathrm{H}(R,C_n) \longrightarrow \mathrm{Disc}(R,n)$.

The inverse δ of d will be constructed explicitly, in a way quite similar to i. Let $(P,\mu) \in \mathrm{Disc}(R,n)$, and define

$$\begin{aligned} R(n;\, P,\mu) &= \mathrm{Sym}_R(P) \big/ \left(\text{ideal generated by all } x - \mu(x),\ x \in P^{\otimes n}\right) \\ &= R \oplus P \oplus \ldots \oplus P^{\otimes (n-1)}, \\ &\qquad \text{with multiplication given by } \mu\colon P^{\otimes n} \to R. \end{aligned}$$

σ acts on $R(n;\, P,\mu)$ by the prescription that it acts as multiplication by ζ on P.

If $P = R$, and μ = multiplication by $u \in R^*$, then $R(n;\, P,\mu)$ is nothing else but $R(n;\, u)$ (see beginning of §). Since we may test the Galois property locally (cf. proof of 4.1), and since each invertible module is locally free, this gives already that $\delta(P,\mu) = R(n;\, P,\mu)$ is indeed a C_n-Galois extension of R.

It is a straightforward calculation to show that d and δ are inverses of each other. It is equally straightforward, and quite similar to the proof for i (5.1. b)) to show that δ is a group homomorphism. Hence d and δ are group isomorphisms, Q.E.D.

From this we shall now deduce the famous Kummer sequence. We want to make it quite clear that our approach here is very old-fashioned, and that the best way of understanding the Kummer sequence is nowadays via cohomology. It has been our wish to keep things very explicit, for later use.

Notation: a) For any commutative ring R, Pic(R) is the set of isomorphism classes of invertible R-modules. Pic(R) is an abelian group, the composition coming from tensor product over R.

b) For any abelian group M, we let as before: $M/n = M/nM$ if M is additive, $M/n = M/M^n$ if M is multiplicative in notation. Similarly, we let $M[n]$ = either $\{x \in M \mid nx = 0\}$, or $\{x \in M \mid x^n = 1\}$, again according to whether M is additive or multiplicative.

Theorem 5.4. *For every n-kummerian ring R, there is a short exact sequence*

$$1 \longrightarrow R^*/n \xrightarrow{i_R} \mathrm{H}(R, C_n) \xrightarrow{\pi} \mathrm{Pic}(R)[n] \longrightarrow 0.$$

The map $i = i_R$ *was defined in* 5.1, *and the map* π *maps* $S \in \mathrm{H}(R, C_n)$ *to the class of the invertible module* $S^{(1)}$.

Proof. There is a canonical map π': Disc$(R,n) \longrightarrow$ Pic(R) mapping (P,μ) to (the class of) P, which is obviously a group homomorphism. The image of π' consists of all invertible modules P such that there exists an isomorphism of $P^{\otimes n}$ with R, i.e. such that the n-th power of P is trivial, hence Im(π') = Pic$(R)[n]$. The kernel of π' consists of all (classes of) discriminant modules (P,μ) with $P \cong R$, i.e. of all classes (R,u), $u \in R^*$. (u is considered as an isomorphism $R^{\otimes n} = R \to R$.) Here (R,u) and (R,v) give the same element of Disc(R,n) iff there exists an isomorphism $R \to R$, i.e. a multiplication by a unit $w \in R^*$, such that the triangle

$$\begin{array}{ccccc} R^{\otimes n} & = & R & \xrightarrow{u} & R \\ & & {\scriptstyle w^n}\big\downarrow & \nearrow_{v} & \\ & & R & & \end{array}$$

commutes, i.e. iff $u = w^n v$ for some $w \in R^*$, whence we get an isomorphism i': $R^*/n \longrightarrow$ Ker(π'), $i'[u]$ = class of (R,u). We obtain a diagram with exact top row

$$\begin{array}{ccccccccc} 1 & \longrightarrow & R^*/n & \xrightarrow{i'} & \mathrm{Disc}(R,n) & \xrightarrow{\pi'} & \mathrm{Pic}(R)[n] & \longrightarrow & 0 \\ & & \| & & \cong\downarrow d & & \| & & \\ 1 & \longrightarrow & R^*/n & \xrightarrow{i} & \mathrm{H}(R,C_n) & \xrightarrow{\pi} & \mathrm{Pic}(R)[n] & \longrightarrow & 0, \end{array}$$

and it is easy to see that this diagram is commutative. This proves the theorem.

Remark: There is a similar result for C_n replaced by an arbitrary finite abelian group G of exponent dividing n.

The Kummer sequence has a companion in characteristic p, so-called Artin-Schreier theory. We will have to treat this in detail in Chapter VI, and the methods employed would lead us too far afield at the moment. Hence we just state the simplest case here. For a proof, see VI §1 or the references given there.

Theorem 5.5. *Let p be any prime and R a ring of characteristic p. Let $\mathbb{P}$: $R \to R$ be the homomorphism of additive groups (!) given by $\mathbb{P}(x) = x^p - x$, $x \in R$. Then there is an isomorphism Φ:*

$$R/\mathbb{P}R \xrightarrow{\Phi} \mathrm{H}(R,C_p)$$

with $j(x) = R[Y]/(Y^p - Y - x)$, the C_p-action being given by $\sigma\overline{Y} = \overline{Y} + 1$.

Remark: In contrast to Kummer theory, there is *no* contribution from Pic(R) in this theory; an explanation for this can be found in the cohomological argument used in VI §1. The Galois operation $\sigma\overline{Y} = \overline{Y}+1$ is an additive analog of the rule $\sigma\overline{X} = \zeta\cdot\overline{X}$ in Kummer theory.

§6 Normal bases and Galois module structure

The material of this section is basic for several chapters of these Notes. Much of it is standard, and may also be found in Chase–Harrison–Rosenberg (1965).

The motivating question is: Given a G–Galois extension S/R (of commutative rings as always), what can be said about the structure of S as an $R[G]$–module? It is clear that the operation of G on S makes S into a left $R[G]$–module. In Galois theory of fields, it is a classical result that for every G–Galois extension L/K, L is free cyclic over $K[G]$, which means in other words that there is a K–basis of L of the form $\{\sigma x \mid x \in G\}$, for some $x \in G$. Such a basis is traditionally called a *normal basis*. Sometimes, the element x by itself is called a normal basis.

We shall see that this result is no longer valid for Galois extensions of rings, but it is very worthwile to study the obstructions. Kummer theory turns out to be quite helpful in doing so. First of all, one has a weakened theorem on normal bases:

Theorem 6.1. *For any G–Galois extension S/R of commutative rings, the $R[G]$–module S is invertible, i.e. finitely generated projective of constant rank* 1.

Proof. The module ${}_RS$ is finitely presented, hence also the $R[G]$–module S (throw in new relations of the form $\sigma{\cdot}x - \sigma(x)$, $\sigma \in G$, x running over a finite system of R–generators of S). Therefore, by descent, it suffices to see that ${}_{R[G]}S$ becomes invertible after a faithfully flat base extension of $R[G]$ (Knus–Ojanguren (1974)). For this extension, we take $S[G] \supset R[G]$. Then $S[G] \otimes_{R[G]} S = S \otimes_R S$ as an $R[G]$–module, and the latter is isomorphic (via h) to $S^{(G)}$, which is obviously free cyclic over $R[G]$ on the element $(1,0\dots,0)$ (the 1 is in position $e \in G$).

Corollary. If R is semilocal, then S is free cyclic over $R[G]$ since $\mathrm{Pic}(R[G]) = 0$.

This result enables us to define a map called "Picard invariant"

$$\mathrm{pic}\colon\ \mathrm{H}(R,G) \longrightarrow \mathrm{Pic}(R[G])$$

(R commutative ring, G finite group), by associating to the class of $S \in \mathrm{H}(R,G)$ the isomorphism class of S as $R[G]$–module. Note that $\mathrm{Pic}(R[-])$ is a covariant functor on *abelian* groups G, a homomorphism $\pi\colon G \to H$ giving rise to the map $P \longmapsto R[H] \otimes_{R[G]} P$. Call this map $\mathrm{Pic}(R,\pi)$.

Proposition 6.2. *On abelian groups,* pic *is a natural transformation, i.e. for every homomorphism π of finite abelian groups, one has a commutative diagram*

$$\begin{array}{ccc} \mathrm{H}(R,G) & \xrightarrow{\text{pic}} & \mathrm{Pic}(R[G]) \\ \pi^* \downarrow & & \downarrow \mathrm{Pic}(R,\pi) \\ \mathrm{H}(R,H) & \xrightarrow{\text{pic}} & \mathrm{Pic}(R[H]) . \end{array}$$

Proof. We can factor π in the form $G \longrightarrow G\times H \longrightarrow H$, the first map being the canonical injection, and the second map being induced by π and id_H. We therefore may do the cases "π is a split mono" and "π is onto" separately. In the case that π is the canonical injection $G \to G\times H$, one has $\pi^*S \cong S\otimes_R R^{(H)}$ for $S \in \mathrm{H}(R,G)$, and $\mathrm{Pic}(R,\pi)(P) = P\otimes_R R[H]$ for $P \in \mathrm{Pic}(R[G])$, hence the diagram commutes (note $R^{(H)} \cong R[H]$). Hence we now assume that π is a canonical epimorphism $G \to G/N = H$.

Let M for a moment be the $R[G]$-module $R[G]$. The following facts are easily checked:

a) M^N (= submodule of elements fixed under N) $= \nu\cdot M$ with $\nu = \Sigma_{\sigma\in N}\sigma$;

b) The annihilator of ν in $R[G]$ is equal to the kernel of $M \longrightarrow R[G/N]\otimes M$ ($\otimes$ over $R[G]$), and these submodules are both equal to the $R[G]$-span of the set $\{1-\sigma \mid \sigma\in N\}$.

By localizing, one sees that a) and b) remain true for every invertible $R[G]$-module M (actually M projective suffices). Choose now $M = S$ (a given element of $\mathrm{H}(R,G)$). Then $\pi^*S = S^N$, and it is our task to prove

$$S^N \cong R[G/N]\otimes_{R[G]} S.$$

Define a map f: $S^N \longrightarrow R[G/N]\otimes_{R[G]} S$ as follows: for $x \in S^N$, pick y in S with $x = \nu y$, and let $f(x) = 1\otimes y$. Properties a) and b) for $M = S$ show that f is well-defined. Since any $y \in S$ can occur, f is surjective. The injectivity of f follows from b). It is also clear that f is $R[G/N]$-linear, q.e.d.

This proposition has an important consequence:

Theorem 6.3. *For G abelian, the map* pic: $\mathrm{H}(R,G) \longrightarrow \mathrm{Pic}(R,G)$ *is a homomorphism. Moreover,* $\mathrm{Pic}(R[G])$ *is in an obvious way a functor in R, and* pic *is natural in R.*

Proof. Let S, $T \in \mathrm{H}(R,G)$. Then $S\otimes_R T$ is a $G\times G$-Galois extension, and it is clear that $\mathrm{pic}_{G\times G}(S\otimes_R T) = \mathrm{pic}(S)\otimes_R \mathrm{pic}(T)$ ($\mathrm{pic}_{G\times G}$ is ad hoc notation for pic: $\mathrm{H}(R,G\times G) \longrightarrow \mathrm{Pic}(R[G\times G])$, and we identify $R[G\times G]$ with $R[G]\otimes_R R[G]$ as usual). We apply $\mathrm{Pic}(R,\mu)$ to the last equation. Consider the right side first. Since $\mathrm{Pic}(R,\mu)$ is just tensoring with $R[G]$ over $R[G\times G]$ via μ, i.e. factoring out modulo the ideal $\mathrm{Ker}(R[G\times G] \to R[G])$, we see that $\mathrm{Pic}(R,\mu)(P\otimes_R Q) \cong P\otimes_{R[G]} Q$ for $P, Q\in \mathrm{Pic}(R[G])$. Ther left side gives, by 6.2, just $\mathrm{pic}(\mu^*(S\otimes_R T))$ which equals $\mathrm{pic}(S\cdot T)$ by definition. This proves that $\mathrm{pic}(S\cdot T) \cong S\otimes_{R[G]} T$, as claimed.

The second statement of the theorem is quite easy to verify.

Definition. A G-Galois extension S/R *has a normal basis*, if S is free cyclic as an $R[G]$-module. (G is a finite group, S and R commutative rings, as always.)

There are the following equivalent formulations of this very important definition:

a) S is free over $R[G]$. (Note that the rank of S over R is uniquely defined.)

b) There exists $x \in S$ such that $\{\sigma x \mid \sigma \in G\}$ is an R-basis of S (such a basis, or sometimes x by itself, is called a normal basis of S over R.)

c) $\mathrm{pic}(S)$ is trivial.

Definition. $\mathrm{NB}(R,G)$ denotes the set of all isomorphism classes of G-Galois extensions S/R which have a normal basis. By definition, $\mathrm{NB}(R,G)$ is a subset of $\mathrm{H}(R,G)$.

Corollary 6.4. *If G is abelian, then* $\mathrm{NB}(R,G)$ *is a subgroup of* $\mathrm{H}(R,G)$.

Proof. Immediate from 6.3 and c) just above.

To simplify notation in the sequel, we introduce another notation:

Definition. For G abelian we define

$$\mathrm{P}(R,G) \quad = \quad \mathrm{H}(R,G)/\mathrm{NB}(R,G).$$

Remark. Of course, $\mathrm{P}(R,G)$ is canonically isomorphic to $\mathrm{Im}(\mathrm{pic}) \subset \mathrm{Pic}(R[G])$. There are several general results on $\mathrm{Im}(\mathrm{pic})$. For instance, $\mathrm{Im}(\mathrm{pic})$ is contained in the subgroup $\mathrm{PrimPic}(R[G]) = \{P \mid \mathrm{Pic}(R,\Delta)(P) = \mathrm{Pic}(R,\iota_1)(P)\cdot\mathrm{Pic}(R,\iota_2)(P)\}$, where Δ, ι_1, ι_2: $G \to G\times G$ are the diagonal, and the two canonical injections, respectively. Cf. Childs (1984). Note that $\mathrm{P}(R,G) = 0$ in case R is semilocal (Cor. to 6.1).

At the end of this section, we describe the connection between Kummer theory (§5) and normal bases.

Proposition 6.5. *Suppose R is n-kummerian ($n \in \mathbb{N}$), and $G = C_n$ (cyclic of order n). Then a C_n-extension S/R has normal basis iff $\pi(S)$ is trivial, where π is the map* $\mathrm{H}(R,C_n) \longrightarrow \mathrm{Pic}(R)[n]$ *in the Kummer sequence 5.4. Consequently, i induces an isomorphism $R^*/n \longrightarrow \mathrm{NB}(R,C_n)$, and π induces an isomorphism* $\mathrm{P}(R,C_n) \longrightarrow \mathrm{Pic}(R)[n]$.

Proof. Suppose first that S has a normal basis, i.e. $S \cong R[G]$ over $R[G]$. Recall that $\pi(S)$ is the (class of the) invertible R-module $S^{(1)} = \{x \in S \mid \sigma x = \zeta\cdot x\}$. ($\sigma$ is a generator of C_n, and ζ a root of Φ_n in R, fixed once and for all.) Hence in our case: $\pi(S) \cong \{y \in R[G] \mid \sigma y = \zeta y\}$, and it is checked at once that the latter R-module is free cyclic, generated by $\sum_{j=0}^{n-1} \zeta^{-j}\sigma^j$. Hence $\pi(S)$ is trivial.

Assume on the other hand that $\pi(S)$ is trivial, i.e. by 5.4: $S \in \mathrm{Im}(i)$, $S = R(n;u)$ for some unit $u \in R$. We explicitly construct a normal basis of S over R: let $\alpha = \overline{X}$ (so $\alpha^n = u$), and let $z = 1+\alpha+\ldots+\alpha^{n-1}$. Then $\sigma^i z = 1+\zeta^i\alpha+\ldots+\zeta^{(n-1)i}\alpha^{n-1}$. Hence the R-linear map f: $S \to S$ given by $\alpha^i \longmapsto \sigma^i z$ ($i = 0,1,\ldots,n-1$), has defining square matrix $A = (\zeta^{ij})$ ($0 \le i,j < n$). We have seen in the proof of 5.1 that this Vandermonde matrix A is invertible in the n-kummerian ring R, hence f is an R-isomorphism, and the $\sigma^i z$ $(i = 0,1,\ldots,n-1)$ are an R-basis of S, i.e. S has a normal basis, q.e.d.

Corollary to this proof: If we let $z = u_0 + u_1\alpha + \ldots + u_{n-1}\alpha^{n-1}$, where $u_0,\ldots,u_{n-1}$ are arbitrary units of R, then $z, \sigma z, \ldots, \sigma^{n-1}z$ is still a normal basis of S, because the representing matrix of $\alpha^i \longmapsto \sigma^i z$ is now $(\zeta^{ij}\cdot u_j)$, which is invertible just as well.

§7 Galois descent

Galois descent is a fragment of the theory of faithfully flat descent. So far, we only have been using a "trivial part" of this technique, useful in testing whether a given map is an isomorphism, say. There is more than that to descent theory. Namely, descent theory is also a means of *constructing* certain morphisms and, most important of all, objects over a ring R which are previously only given over a faithfully flat extension S. Briefly, one wants to solve the equation $S \otimes_R X \cong Y$ for X. (X might be an R-module, an R-algebra, . . .) This section is about this constructive part of descent theory, limited to the case of *Galois descent*, i.e. S/R a Galois extension. The results will be used frequently in later chapters.

The motivating question, therefore, is: Given a G-Galois extension S/R (of commutative rings), and an S-module (or: S-algebra) N, when is N, up to isomorphism, of the form $S \otimes_R M$, for an R-module M (or: R-algebra M)? If ψ: $S \otimes_R M \longrightarrow S \otimes_R M'$ is a homomorphism of S-modules (S-algebras), when does ψ have the form $S \otimes \varphi$? Galois descent gives a complete answer to both questions; the second one is slightly easier.

Convention. To avoid repetition, the word "R-object" in this section is supposed to mean consistently *either* "R-module" *or* "R-algebra" *or* "R-algebra with action of a given group C by R-automorphisms". All statements are meant simultaneously for these three kinds of objects. Accordingly, the word "R-morphism" means either "R-module hom." or "R-algebra hom." or "C-equivariant R-algebra hom." There exists of course a categorical framework encompassing all these cases and much more, but for our purposes a more down-to-earth approach is preferable. For the general theory, the reader may consult Grothendieck (1959).

Definition. Let S/R be a G-Galois extension.

a) An S-morphism f: $A \to B$ between two S-objects is called σ-*linear* (for some $\sigma \in G$), if $f(sa) = \sigma(s)f(a)$ for all $s \in S$, $a \in A$.

b) A *descent datum* $\Phi = (\Phi_\sigma)_{\sigma\in G}$ on some S-object B is a family of R-automorphisms Φ_σ of B such that: Φ_σ is σ-linear for all $\sigma \in G$, and $\Phi_\sigma\Phi_\tau = \Phi_{\sigma\tau}$ for all $\sigma, \tau \in G$.

Example. If $B = S \otimes_R A$ for some R-object A, then there is the so-called *trivial descent datum* $(\Phi_\sigma)_{\sigma\in G}$ defined by $\Phi_\sigma(s \otimes a) = \sigma(s) \otimes a$ for $s \in S$, $a \in A$.

Theorem 7.1. *Let B be an S-object, and Φ a descent datum on it. Then $A = B^\Phi$ (which equals by definition $\{b \in B \mid \Phi_\sigma(b) = b$ for all $\sigma \in G\}$) is an R-object; the canonical map α: $S \otimes_R A \longrightarrow B$ induced by $A \to B$ is an isomorphism, and the trivial descent datum on $S \otimes_R A$ corresponds via α to the given descent datum Φ.*

Proof. It is immediate that A is indeed an R-object (recall the above convention). The point is to show that α is bijective. The last statement of the theorem is then a direct consequence.

Let T be any faithfully flat R-algebra such that S_T $(= T \otimes_R S)$ is the trivial G-extension of T. ($T = S$ is a possibility.) It then suffices to show that α_T: $S_T \otimes_{R_T} A_T \longrightarrow B_T$ is an isomorphism. Now since tensoring with T preserves kernels, A_T is precisely the fixed ring of all $T \otimes \Phi_\sigma$, $\sigma \in G$. Hence we may change notation and assume to begin with: S/R is the trivial G-extension: $S = R^{(G)} = \bigoplus_{\sigma\in G} e_\sigma \cdot R$, where the e_σ are the standard idempotents, and G acts via $\tau{*}e_\sigma = e_{\sigma\tau^{-1}}$ $(\sigma, \tau \in G)$. Then B likewise splits in the form $B = \bigoplus_{\sigma\in G} B_\sigma$ with $B_\sigma = e_\sigma B$, and each B_σ is an R-object. One checks that the τ-linear automorphism Φ_τ of B is given by a family $(f_\sigma^{(\tau)})_{\sigma\in G}$ of isomorphisms $f_\sigma^{(\tau)}$: $B_\sigma \to B_{\sigma\tau^{-1}}$, and these isomorphisms satisfy the condition

$$f_{\sigma\tau^{-1}}^{(\rho)} f_\sigma^{(\tau)} = f_\sigma^{(\rho\tau)} \quad (\sigma, \tau, \rho \in G).$$

Therefore the B_σ are canonically isomorphic to one R-object A' (take e.g. $A' = B_1$), so B becomes identified with $\bigoplus_{\sigma\in G} A'$, and the descent datum Φ now operates just by index shift. One then obtains $A = \mathrm{diag}(A') \subset A'^{(G)} = B$, and it is now obvious that α: $R^{(G)} \otimes_R A \longrightarrow B$ is an isomorphism, q.e.d.

One also needs to descend morphisms. This works as follows.

Proposition 7.2. *Let A_1, A_2 be two R-objects, g: $S \otimes_R A_1 \longrightarrow S \otimes_R A_2$ an S-morphism, and $(\Phi_\sigma^{(i)})_{\sigma\in G}$ the trivial descent datum on $S \otimes_R A_i$ $(i = 1,2)$. Then g is of the form $S \otimes f$ for some R-morphism f: $A_1 \to A_2$ iff $\Phi_\sigma^{(2)} g = g\Phi_\sigma^{(1)}$ for all $\sigma \in G$. Moreover, f is unique if it exists.*

Proof. The uniqueness of f is immediate from the fact that S is faithfully flat over R.

If $g = S \otimes f$, then one checks the formula $\Phi_\sigma^{(2)} g = g\Phi_\sigma^{(1)}$ directly, just using the definitions. The other implication is shown as follows: A_2 embeds into $S \otimes_R A_2$ by faithful flatness. Moreover, A_2 is contained in the fixed object A of all $\Phi_\sigma^{(2)}$,

$\sigma \in G$. If A were strictly larger than A_2, then the inclusion ι: $A_2 \to A$ would not be an isomorphism, contradicting the fact that $S \otimes \iota$: $S \otimes_R A_2 \longrightarrow S \otimes_R A = S \otimes_R A_2$ is an isomorphism by 7.1. Since A_1 is fixed under all $\Phi_\sigma^{(1)}$, our hypothesis implies that $g(A_1)$ is fixed under all $\Phi_\sigma^{(2)}$, hence contained in A_2. Let $f = g|A_1$. This gives a well-defined R-morphism f: $A_1 \to A_2$, and $S \otimes f = g$.

As an application, we prove:

Proposition 7.3. *Let S/R be a G-Galois extension of* <u>*connected*</u> *rings. Then* $\mathrm{Aut}(S/R)$ *is equal to G.*

Proof. We shall use the S-isomorphism h: $S \otimes_R S \longrightarrow S^{(G)}$. Now $S \otimes_R S$ carries the trivial descent datum $(\Phi_\sigma) = (\sigma \otimes \mathrm{id}_S)_{\sigma \in G}$, and descending along (Φ_σ) gives back S, as we have seen in the proof of 7.2. Let us transport (Φ_σ) via h to a decent datum (Φ_σ') on $S^{(G)}$. A short calculation shows $\Phi_\sigma'\big((x_\tau)_{\tau \in G}\big) = (\sigma(x_{\sigma^{-1}\tau}))_{\tau \in G}$.

By 7.2, the R-automorphisms f of S correspond bijectively to the S-automorphisms g of $S^{(G)}$ that commute with all Φ_σ'. Using that S is connected, one sees that every S-automorphism g of $S^{(G)}$ is given by some permutation α of G in the obvious way $(g(x_\tau)_\tau = (x_{\alpha(\tau)})_\tau)$. Moreover, such a g commutes with Φ_σ iff α commutes with left multiplication by σ^{-1}. Hence g descends to an R-automorphism f iff the permutation α is right multiplication with some element of G, but this already means that g is in G.

§8 $\mathbb{Z}_p$-extensions

To conclude this introductory chapter, we briefly discuss extensions with "Galois group" $\mathbb{Z}_p$, the additive group of p-adic numbers, or, if one prefers, the inverse limit of $\mathbb{Z}/p^n$ for $n \to \infty$. Such extensions are best thought of as *towers* (i.e., sequences satisfying a certain compatibility) of extensions with Galois group C_{p^n} = (cyclic group of order p^n), one for each $n \in \mathbb{N}$.

Fix a prime number p, and in each C_{p^n} fix a generator σ_n. Define epimorphisms π_n: $C_{p^{n+1}} \to C_{p^n}$ by setting $\pi_n(\sigma_{n+1}) = \sigma_n$. The groups C_{p^n} together with the π_n form a projective (or "inverse") system. The kernel of π_n is the unique subgroup of $C_{p^{n+1}}$ of order p. The projective limit of the system (C_{p^n}, π_n) is canonically isomorphic to $\mathbb{Z}_p$, the sequence $(\sigma_n)_{n \in \mathbb{N}}$ corresponding to $1 \in \mathbb{Z}_p$.

Definition. Let R be a commutative ring. The *group* $\mathrm{H}(R, \mathbb{Z}_p)$ *of $\mathbb{Z}_p$-extensions of R* is defined as the projective limit of the system

$$\cdots \xrightarrow{\pi_n^*} \mathrm{H}(R, C_{p^n}) \longrightarrow \cdots \xrightarrow{\pi_2^*} \mathrm{H}(R, C_{p^2}) \xrightarrow{\pi_1^*} \mathrm{H}(R, C_p).$$

Thus, the elements of $\mathrm{H}(R,\mathbb{Z}_p)$ are sequences $(A_n)_{n\in\mathbb{N}}$ with $A_n \in \mathrm{H}(R,C_{p^n})$ and $\pi_n^*(A_{n+1}) = A_n$ for all $n\in\mathbb{N}$. Such sequences are also called *coherent*, or *towers*.

Remark. Since $\pi_n^*(A_{n+1})$ is just the fixed ring of $\mathrm{Ker}(\pi_n)$ in A_{n+1} (because π_n is onto!), one may regard A_n as a subring of A_{n+1} for all n, and it makes sense to talk about the R-algebra

$$A_\infty = \bigcup_{n=1}^{\infty} A_n.$$

Now $\varprojlim C_{p^n}$ ($\cong \mathbb{Z}_p$) is a pro-p-group with topological generator $\sigma_\infty = (\ldots,\sigma_2,\sigma_1)$, and there is a canonical action of this pro-p-group on A_∞, as follows: for $\tau \in \varprojlim C_{p^n}$ and $x \in A_\infty$, find $n \in \mathbb{N}$ with $x \in A_n$, and let τ_n be the image of τ in C_{p^n}. Define $\tau(x)$ to be $\tau_n(x)$. This is independent of the choice of n, thanks to the compatibility condition $\pi_n^*(A_{n+1}) = A_n$. (If A_∞ is connected, one can even show that $\mathrm{Aut}(A_\infty/R) \cong \varprojlim C_{p^n}$, cf. 7.3.) This justifies the terminology "$\mathbb{Z}_p$-extension".

In more succinct notation, the definition reads $\mathrm{H}(R,\mathbb{Z}_p) = \varprojlim_n \mathrm{H}(R,C_{p^n})$.

Definition. The *group* $\mathrm{NB}(R,\mathbb{Z}_p)$ *of* $\mathbb{Z}_p$*-extension of* R *with normal basis* is defined by

$$\mathrm{NB}(R,\mathbb{Z}_p) = \varprojlim_n \mathrm{NB}(R,C_{p^n}).$$

(Recall in this context that π_n^* is just as well a homomorphisms $\mathrm{NB}(R,C_{p^{n+1}}) \to \mathrm{NB}(R,C_{p^n})$ for all $n \in \mathbb{N}$.)

Since projective limits preserve monomorphisms, the group $\mathrm{NB}(R,\mathbb{Z}_p)$ is a subgroup of $\mathrm{H}(R,\mathbb{Z}_p)$. We thus may define $\mathrm{P}(R,\mathbb{Z}_p) = \mathrm{H}(R,\mathbb{Z}_p)/\mathrm{NB}(R,\mathbb{Z}_p)$. Caution: it is not clear whether $\mathrm{P}(R,\mathbb{Z}_p) \cong \varprojlim \mathrm{P}(R,C_{p^n})$. If all groups $\mathrm{NB}(R,C_{p^n})$ happen to be finite, then this isomorphism does hold, since the derived functor $\varprojlim^{(1)}$ is zero on $(\mathrm{NB}(R,C_{p^n}))_n$. Cf. Jensen (1972).

The topology on $\mathbb{Z}_p \cong \varprojlim C_{p^n}$ is the profinite topology, and induced from the product topology on $\prod_{n\in\mathbb{N}} C_{p^n}$. In the context of §3, we then have the following result:

Lemma 8.1. *For connected rings* R, *there is a canonical isomorphism* $\mathrm{H}(R,\mathbb{Z}_p) \longrightarrow \mathrm{Hom}_{cont}(\Omega_R,\mathbb{Z}_p)$ *(which is the same as* $\mathrm{Hom}_{cont}(\Psi_R,\mathbb{Z}_p)$ *by Cor. 3.9).*

Proof.

$$\begin{aligned} \mathrm{H}(R,\mathbb{Z}_p) &\cong \varprojlim \mathrm{H}(R,C_{p^n}) \\ &\cong \varprojlim \mathrm{Hom}_{cont}(\Omega_R,C_{p^n}) \quad \text{(by Thm. 3.5)} \\ &\cong \mathrm{Hom}_{cont}(\Omega_R,\varprojlim C_{p^n}) \quad \text{(universal property of } \varprojlim) \\ &\cong \mathrm{Hom}_{cont}(\Omega_R,\mathbb{Z}_p). \end{aligned}$$

CHAPTER I

Cyclotomic descent

This chapter lays the general technical foundations for later chapters, in which the group $\mathrm{H}(R, C_{p^n})$ of cyclic Galois extensions of degree p^n is studied in more detail for specific rings R. Recall that C_{p^n} is the cyclic group of order p^n.

The basic idea is very simple. If R is p^n-kummerian, then $\mathrm{H}(R, C_{p^n})$ is described by Kummer theory, and there is not much more to say. For arbitrary R, one therefore tries to pass to some p^n-kummerian overring $R_n \supset R$ and to obtain results for R_n, in the hope of being able to get back to R somehow. Experience shows that this works smoothly only if $p^{-1} \in R$. If this is the case, then one may even choose R_n to be Galois over R, with well-understood Galois group, and the techniques of Galois descent (**0** §7) apply. There is a sort of minimal choice of R_n, and this will be denoted S_n. This construction will occupy §1, and then the actual descent machinery will be developed.

Another important matter in this chapter is to understand the behavior of normal bases under the kind of descent just sketched.

§1 Cyclotomic extensions

Fix a prime p and a natural number $n \geq 1$. Let R be any connected (commutative!) ring R in which p is invertible. The p^nth cyclotomic polynomial

$$\Phi_n = X^{p^{n-1}(p-1)} + X^{p^{n-1}(p-2)} + \ldots + X^0 = \frac{X^{p^n}-1}{X^{p^{n-1}}-1}$$

lies in $\mathbb{Z}[X]$, and we shall also consider it as an element of $R[X]$. Note that we write Φ_n instead of Φ_{p^n}, to save subscripts.

It is well-known that Φ_n is irreducible over $\mathbb{Q}$. Let ζ_n be a primitive p^n-th root of unity over $\mathbb{Q}$, or, what is the same, a root of Φ_n. (Note again the index simplification.) Then $\mathbb{Q}_n = \mathbb{Q}(\zeta_n)$ is isomorphic to $\mathbb{Q}[X]/(\Phi_n(X))$, and $\mathbb{Q}_n$ is a Galois field extension of $\mathbb{Q}$, because $\Phi_n/\mathbb{Q}$ is irreducible and splits completely in $\mathbb{Q}_n$ (recall the formula $\Phi_n = \prod_i (X - \zeta_n^i)$ where i runs over all integers from 0 to p^n which are coprime to p). Let Δ_n be the Galois group of $\mathbb{Q}_n/\mathbb{Q}$. Then there is a natural group homomorphism $\omega\colon \Delta_n \longrightarrow (\mathbb{Z}/p^n)^*$ (the unit group of the ring $\mathbb{Z}/p^n\mathbb{Z}$), which is given by the prescription $\tau(\zeta_n) = \zeta_n^{\omega(\tau)}$ for all $\tau \in \Delta_n$. (Note that $\tau(\zeta_n)$ is certainly again a primitive p^n-th root of unity, hence of the form ζ_n^w for some w coprime to p.) Since any $\tau \in \Delta_n$ is determined by its value on ζ_n, the map ω is injective, and ω is obviously a homomorphism. Since $|\Delta_n| = \deg(\Phi_n) = p^{n-1}(p-1) = |(\mathbb{Z}/p^n)^*|$, we get that ω is even bijective, and consequently $\Delta_n \cong (\mathbb{Z}/p^n)^*$.

Let $\mathbb{Z}' = \mathbb{Z}[p^{-1}]$, $\mathbb{Z}_n' = \mathbb{Z}[p^{-1}, \zeta_n] \subset \mathbb{Q}_n$.

Lemma 1.1. *$\mathbb{Z}_n'$ is a Δ_n-Galois extension of $\mathbb{Z}'$.*

Proof. We give two proofs. If one is willing to use some algebraic number theory, one may argue as follows: $\mathbb{Z}[\zeta_n]$ is the ring of integers in $\mathbb{Q}_n$ (see e.g. Washington (1982), Ex. 1.1), hence $\mathbb{Z}[\zeta_n, p^{-1}]$ is the ring of p-integers in $\mathbb{Q}_n$, i.e. the ring of elements of $\mathbb{Q}_n$ which have nonnegative valuation at all primes not dividing p; note here that $\mathbb{Q}_n$ has only one prime ideal dividing p, namely $(1-\zeta_n)$. Moreover $\mathbb{Q}_n/\mathbb{Q}$ is unramified outside p (loc. cit. Prop. 2.3), hence by Thm. 0.4.1, $\mathbb{Z}_n' = \mathbb{Z}[\zeta_n, p^{-1}]$ is Δ_n-Galois over $\mathbb{Z}'$, the ring of p-integers in $\mathbb{Q}$.

We also give a direct argument from first principles: Look at the map h: $\mathbb{Z}_n' \otimes_{\mathbb{Z}'} \mathbb{Z}_n' \longrightarrow \mathbb{Z}_n'^{(\Delta_n)}$; we want to show it is an isomorphism. Inserting at the second occurence of $\mathbb{Z}_n'$ the isomorphism $\mathbb{Z}_n' \cong \mathbb{Z}'[X]/(\Phi_n)$, $\zeta_n \longmapsto$ residue class of X, we obtain a homomorphism

$$\begin{aligned} h'\colon \mathbb{Z}_n'[X]/\big(\Phi_n(X)\big) &\longrightarrow \mathbb{Z}_n'^{(\Delta_n)}, \\ \bar{X} &\longmapsto (\tau(\zeta_n))_{\tau\in\Delta_n}. \end{aligned}$$

Now $\Phi_n(X) = \prod_{\tau\in\Delta_n}(X - \tau(\zeta_n))$ in $\mathbb{Z}_n'[X]$, and h' is just the composite map

$$\mathbb{Z}_n'[X]/\big(\Phi_n(X)\big) \xrightarrow{can} \prod_{\tau\in\Delta_n}\mathbb{Z}_n'[X]/(X-\tau(\zeta_n)) = \prod_{\tau\in\Delta_n}\mathbb{Z}_n' = \mathbb{Z}_n'^{(\Delta_n)},$$

so we are done if the map labeled *can* is an isomorphism. But this follows from the Chinese Remainder Theorem as soon as we know that the polynomials $X - \tau(\zeta_n)$ are coprime in pairs in $\mathbb{Z}_n'[X]$. This, in turn, is equivalent to the fact that all differences $\zeta_n^i - \zeta_n^j$ with $i \neq j$, $i, j \in (\mathbb{Z}/p^n)^*$, are units in $\mathbb{Z}_n'$. For this, it is enough to show that $1-\zeta_n^s$ is a unit for all $0 < s < p^n$. This is a well-known fact (one proof: $1-\zeta_n^s$ divides $\prod_{0<t<p^n}(1-\zeta_n^t) = \left(\frac{X^{p^n}-1}{X-1}\right)\Big|_{X=1} = p^n$, which is a unit in $\mathbb{Z}_n'$.) Q.E.D.

By base extension with R over $\mathbb{Z}' = \mathbb{Z}[p^{-1}]$ (recall R is any commutative ring containing p^{-1}) we get at once:

Corollary 1.2. *$R_n = R[X]/(\Phi_n(X))$ is a Δ_n-Galois extension of R, with Δ_n-action given by $\tau(X) = \bar{X}^{\omega(\tau)}$ for $\tau \in \Delta_n$. Moreover, R_n is p^n-kummerian.*

There is the technical problem that a *connected* p^n-kummerian Galois extension of R would be preferable, especially in number theory. (In general R_n, as defined in the last corollary, is not connected.) By Prop. **0**.3.8, there is a subgroup $\Gamma_n \subset \Delta_n$ (depending on R) and a connected Γ_n-extension S_n/R such that $R_n \cong \iota^* S_n$, where $\iota\colon \Gamma_n \subset \Delta_n$ is the inclusion.

Lemma 1.3. *a) S_n is p^n-kummerian and connected, generated by a root of Φ_n as an R-algebra.*

b) Γ_n is unique, i.e. if $R_n \cong \eta^* T_n$, where $\eta\colon \Gamma' \subset \Delta_n$ is another inclusion of a subgroup, and T_n/R is Γ'-Galois, then $\Gamma_n = \Gamma'$.

c) The class of S_n in $\mathrm{H}(R,\Gamma_n)$ is uniquely determined.

Proof. From the definition of ι^*, one has $R_n = \mathrm{Map}_{\Gamma_n}(\Delta_n, S_n)$. As an R-algebra, this is isomorphic to a direct product $(S_n)^{(\Delta_n/\Gamma_n)}$, and Δ_n/Γ_n acts simply transitively on the set of indecomposable idempotents of this algebra (recall S_n has no nontrivial idempotents). Hence Γ_n may be characterized intrinsically as the stabilizer of any indecomposable idempotent of R_n, which proves b). Part a) is obvious since S_n is an epimorphic image of the R-algebra R_n. To see c), note that S_n has the following intrinsic description: take any indecomposable idempotent e of R_n, and let $S_n = eR_n$, as algebra and Γ_n-set.

Definition. The Γ_n-Galois extension S_n of R just constructed is called *the p^n-th cyclotomic extension of R.*

There is another characterization of S_n:

Lemma 1.4. *If S/R is separable, connected, and generated by a root z of Φ_n as an R-algebra, then $S \cong S_n$ as R-algebras. In particular, S/R is Γ_n-Galois, and the action of Γ_n on z is described by the embedding $\omega\colon \Gamma_n \to (\mathbb{Z}/p^n)^*$ via the formula $\tau(z) = z^{\omega(\tau)}$, $\tau \in \Gamma_n$.*

Proof. There is an R-epimorphism $\alpha\colon R_n \to S$ sending $\overline{X}$ to z. Since R_n and S are both separable over R, they both become isomorphic to a product of copies of the ground ring after some faithfully flat base extension, hence by the addendum to **0**.3.7, the kernel of α is generated by an idempotent. Moreover, R_n is isomorphic to a product of copies of S_n, and S_n and S are connected. Hence α induces an isomorphism between S_n and S, sending the image of X in S_n to z. This proves $S \cong S_n$, and also the statement about the action of Γ_n.

Corollary. *If S/R is connected, generated by a root of Φ_n, and Γ-Galois for some finite group Γ, then S is separable, hence by 1.4, $S \cong S_n$, and also $\Gamma \cong \Gamma_n$, since the Galois group of S/R is unique by* **0**.7.3.

This corollary allows us to rephrase the last definition: S_n is *the* Galois extension of R which is connected and generated by a root of Φ_n.

Our next job is to compare the S_n for various n. We first need another definition and a lemma.

Definition. $\mu_{p^n}(R) = \{x \in R \mid x^{p^n}=1\}$. (This makes sense for any ring R.)

Lemma 1.5. *Recall our standing hypothesis $p^{-1} \in R$.*

a) If R is connected and contains a root z of Φ_n, then $\mu_{p^n}(R)$ is the multiplicative group generated by z, and has exactly p^n elements.

b) If R is just connected, then the set of roots of Φ_n in R is $\mu_{p^n}(R) - \mu_{p^{n-1}}(R)$.

Proof. a) Since Φ_n divides $X^{p^n}-1$, the order of z divides p^n. Suppose it were less than p^n, i.e. suppose the p^{n-1}-th power of z were 1. Then $\Phi_n(z) = 1+\ldots+1 = p$, contradicting $\Phi_n(z) = 0$.

Now suppose $y \in \mu_{p^n}(R)$. For $0 \le i < p^n$, let $e_i = p^{-n}\cdot\sum_{j=0}^{p^n-1} y^j\cdot z^{-ij}$. One checks using $\Phi_n(z) = 0$ that for $0 < i < p^n$, $\sum_{j=0}^{p^n-1} z^{-ij} = 0$, and that the e_i are orthogonal idempotents with sum 1. One also checks that $y = \sum_{j=0}^{p^n-1} e_i z^i$. Now since R is connected, exactly one e_k is 1, and all other e_i are zero. Therefore $y = z^k$, q.e.d.

b) "$\subset$": By the proof of a), every root of Φ_n in R (if any exist) has multiplicative order precisely p^n.

"$\supset$": Let y be in the right hand side, i.e. y has order exactly p^n. Then $w = y^{p^n}$ has multiplicative order exactly p. Let $e = p^{-1}(1+w+\ldots+w^{p-1})$. Then $e^2 = e$, hence e is 0 or 1. The case $e = 1$ is impossible since $(1-w)\cdot e = 0$ but $1-w \neq 0$. Hence $e = 0$, which just says that $\Phi_n(y) = 0$, q.e.d.

Proposition 1.6. *Let $m \ge n \ge 1$ be natural numbers.*

a) There is an embedding $S_n \to S_m$ of R-algebras. This embedding is an equality iff S_n contains already a root of Φ_m.

b) There is a canonical surjection $\pi\colon \Gamma_m \to \Gamma_n$, and S_n is just the fixed ring of Ker(π) *in S_m. Moreover, π is compatible with the embeddings $\omega_n\colon \Gamma_n \to (\mathbb{Z}/p^n)^*$ and $\omega_m\colon \Gamma_m \to (\mathbb{Z}/p^m)^*$.*

c) The rank $[S_m{:}S_n]$ divides p^{m-n}. In the case of equality, we have

$$S_m \cong S_n[Y]/(Y^{p^{m-n}} - z), \quad \text{with } z \in S_n \text{ a root of } \Phi_n.$$

Proof. a) Let S_m' be the p^m-th cyclotomic extension of S_n (*not* of R). Then $S_m' = S_n[y]$, y a root of Φ_m. We also have $S_n = R[z]$, z a root of Φ_n. Then by 1.5, y generates the cyclic group $\mu_{p^m}(S_m')$ of order p^m, and z is a p^n-th root of unity, hence z is a power of y. Therefore S_m' is generated by y over R, and separable over R (transitivity of separability). By 1.4, we obtain $S_m' \cong S_m$, hence S_n embeds into S_m'. If $S_n \cong S_m$, then of course S_n contains a root of Φ_m. If, conversely, $y \in S_n$ is such a root, then S_m' equals S_n by 1.4, hence $S_m \cong S_n$.

b) Since the automorphism groups of S_m (S_n) are precisely Γ_m (Γ_n resp.) by **0**.7.3, this follows from the Main Theorem of Galois Theory **0**.2.3. The second statement is easily checked, using how the Galois groups operate on a root y of Φ_m, and the fact that all roots of Φ_n are powers of y.

c) The kernel of π is via ω_m isomorphic to some subgroup of the kernel of the natural map $(\mathbb{Z}/p^m)^* \longrightarrow (\mathbb{Z}/p^n)^*$, and this kernel has order p^{m-n}. This proves the first statement. For the second, suppose $[S_m:S_n] = p^{m-n}$ and define $T = S_n[Y]/(Y^{p^{m-n}}-z)$ with $z \in S_n$ some fixed root of Φ_n. Then $y = \overline{Y}$ is a root of Φ_m and generates T over R. Moreover, T/R is separable. (Proof: It suffices to see T/S_n separable, and this one may see either by direct calculation, or by base-extending with S_{m-n} in case $m-n < n$; by doing so, one even obtains a Kummer extension, which is Galois, hence separable.) We take an indecomposable idempotent e of T; then by 1.4, we have $eT \cong S_m$, hence $[eT:R] = [S_m:R] = [S_m:S_n][S_n:R] = p^{m-n}[S_n:R] = [T:R]$, which is only possible for $e = 1$, hence $T \cong S_m$ as claimed.

Let us now introduce another piece of notation.

Definition. Recall that we are assuming R connected and $p^{-1} \in R$. Define $n_0(R)$ by

$n_0(R) = \infty$ if Φ_n has a root in S_1 for all $n > 1$;

$n_0(R) =$ the largest n such that Φ_n has a root in S_1, otherwise.

Example: $n_0(\mathbb{Z}[p^{-1}]) = 1$; this follows by considering the degrees $[\mathbb{Q}(\zeta_n):\mathbb{Q}]$.

Remark. a) By looking at field degrees as in the example, one can show that $n_0(R)$ is finite for every subring R of a number field.

b) By Lemma 1.5, $p^{n_0(R)}$ is the order of the group $\mu_{p^\infty}(S_1)$, where we have put $\mu_{p^\infty}(S_1) = \bigcup_{n\geq 1} \mu_{p^n}(S_1)$.

Proposition 1.7. *Suppose $p \neq 2$, and let $n_0 = n_0(R)$.*

a) For $1 \leq n \leq n_0$, S_n is canonically isomorphic to S_1.

b) For $n \geq n_0$, we have $[S_{n+1}:S_n] = \mathrm{rank}_{S_n}(S_{n+1}) = p$, and $\mu_{p^\infty}(S_n) = \mu_{p^n}(S_n)$.

Proof. We have seen in 1.6 that S_{n+1}/S_n is D-Galois, with $D = \mathrm{Ker}(\Gamma_{n+1} \to \Gamma_n)$, and that $|D| = 1$ or p, according to whether or not Φ_{n+1} has a root in S_n or not. Taking $n = 1,\dots,n_0-1$, we get $S_1 \cong \dots \cong S_{n_0}$. Taking $n = n_0$, we get the first assertion of b) for $n = n_0$. We show $[S_{n+1}:S_n] = p$ by induction over $n \geq n_0$, the case $n = n_0$ being settled. The kernel D is embedded via ω_{n+1} in D', the kernel of $(\mathbb{Z}/p^{n+1})^* \longrightarrow (\mathbb{Z}/p^n)^*$, and D' is the group of order p generated by the class of $1+p^n$ mod p^{n+1}. Hence $[S_{n+1}:S_n] = p$ iff $\omega_{n+1}(\Gamma_{n+1})$ contains $1+p^n$. Thus, suppose $\omega_{n+1}(\Gamma_{n+1})$ contains $1+p^n$; we must show $\omega_{n+2}(\Gamma_{n+2})$ contains $1+p^{n+1}$. By 1.6 b), $\Gamma_{n+2} \to \Gamma_{n+1}$ is onto. Let $u = \overline{1+p^n+ap^{n+1}}$ $(0 \leq a < p-1)$ be a preimage of $\overline{1+p^n}$ in $\omega_{n+2}(\Gamma_{n+2})$. Using $p \neq 2$, one sees easily that $u^p = \overline{1+p^{n+1}} \in \mathbb{Z}/p^{n+2}$.

The second statement in b) is a consequence of the first one and 1.6 a).

Remark. For $p = 2$, one has to replace S_1 by S_2 in the definition of $n_0(R)$, and in Prop. 1.7.

At a later occasion, we shall need the so-called p-part of the cyclotomic extension S_n/R. Suppose p *odd* for the rest of the §. Since the group $(\mathbb{Z}/p^n)^*$ is cyclic, also Γ_n is cyclic for all $n \geq 1$. We therefore have a decomposition $\Gamma_n = \Gamma_n' \oplus \Gamma_n''$, with Γ_n' a (cyclic) p-group, and the order of the group Γ_n'' prime to p. By 1.7 we have that $|\Gamma_n| = |\Gamma_1|$ for $n \leq n_0$, and $|\Gamma_n| = p^{n-n_0} \cdot |\Gamma_{n_0}|$ for $n > n_0$. Moreover, $|\Gamma_1|$ is prime to p since it divides $|(\mathbb{Z}/p)^*| = p-1$. Hence the order of Γ_n' is 1 for $n \leq n_0$ and p^{n-n_0} for $n \geq n_0$.

Let now A_n be the fixed ring of Γ_n'' in S_n. By the main theorem (**0** §2), A_n is then a Γ_n'-Galois extension of R. As we have just seen, Γ_n' is cyclic of order p^{n-n_0} for $n \geq n_0$. The inclusions $S_n \subset S_{n+1}$ give also inclusions $A_n \subset A_{n+1}$, and the resulting maps $\Gamma_{n+1}' \longrightarrow \Gamma_n'$ are onto.

It is therefore reasonable to identify Γ_{n+n_0}' with C_{p^n}, the cyclic group of order p^n with generator σ_n. This identification is not quite canonical. However, it is not hard to see that the only subgroup of order p^n in $(\mathbb{Z}/p^{n+n_0})^*$ is generated by the class of $1+p^{n_0}$, and one may identify $\omega^{-1}(\overline{1+p^{n_0}})$ with σ_n. This is a good choice, since the map $\Gamma_{n+1}' \to \Gamma_n'$ then is given by $\sigma_{n+1} \longmapsto \sigma_n$. With this identification, $A_{n+n_0} \in \mathrm{H}(R, C_{p^n})$ for all $n \geq 0$. Let us abbreviate $A_{n+n_0} = B_n$ and call it the *n-th cyclotomic p-extension of R*. (Caution: S_{n+n_0} is called the p^{n+n_0}-th cyclotomic extension!)

We shall need to know the subgroup of $\mathrm{H}(R, C_{p^n})$ generated by B_n. The answer is given by the following more general lemma:

Lemma 1.8. *Let A/R be a connected C_{p^n}-Galois extension. Then $A \in \mathrm{H}(R, C_{p^n})$ generates a subgroup of order p^n.*

Proof. Think of C_{p^n} as an additive group. We have to show: $m \cdot A$ (= m-fold Harrison product of A with itself) is trivial iff p^n divides m. Let $\mu_m\colon C_{p^n}{}^m \longrightarrow C_{p^n}$ be the sum map, and $\iota_m\colon C_{p^n} \longrightarrow C_{p^n}{}^m$ the diagonal. Then

$$\begin{aligned} m \cdot A &= \mu_m^*(A \otimes_R \cdots \otimes_R A) && (m \text{ factors}) \\ &= \mu_m^* \iota_m^*(A) && (\text{cf. proof of } \mathbf{0}.3.2 \text{ a)}). \end{aligned}$$

Now $\mu_m \iota_m = [m]$, the multiplication by m from C_{p^n} to C_{p^n}, hence $m \cdot A = [m]^* A$. If $[m]^* A$ is trivial, then by **0**.3.3, $A^{\mathrm{Ker}([m])}$ is trivial (as a Coker($[m]$)-Galois extension of R). Since A is connected, this is possible only for $[A^{\mathrm{Ker}([m])}:R] = 1$, i.e. $\mathrm{Ker}([m]) = C_{p^n}$, which is equivalent to m being a multiple of p^n. (Cf. second remark following Thm. **0**.3.5.)

§2 Descent of normal bases

In this section we fix a ground ring R which is connected and contains p^{-1}. In the preceding § we constructed the p^n-th cyclotomic extension S_n of R and discussed its properties.

Definition. j_n: $H(R, C_{p^n}) \longrightarrow H(S_n, C_{p^n})$ is the group homomorphism given by $j_n([A]) = [S_n \otimes_R A]$ (base extension with S_n over R).

Remark. By Cor. **0**.6.4, j_n restricts to a homomorphism $NB(R, C_{p^n}) \to NB(S_n, C_{p^n})$. (The reason is simply that whenever x generates a normal basis of A over R, then $1 \otimes x$ generates a normal basis of $S_n \otimes_R A$ over S_n.)

Theorem 2.1. *If $p \neq 2$ or R contains a primitive 4th root of unity (i.e. a root of X^2+1), then*

$$j_n^{-1}(NB(S_n, C_{p^n})) = NB(R, C_{p^n}).$$

In words: A C_{p^n}-extension of R has a normal basis if and only it has a normal basis after base extension with S_n.

Proof. "$\supset$" follows from the above remark.

For the proof of the other inclusion (which is nontrivial), let us start with $A \in H(R, C_{p^n})$ such that $A' = S_n \otimes_R A$ has a normal basis over S_n. The explanation in the above remark can be complemented as follows: if $y \in A$, and if $1 \otimes y$ generates a normal basis of A', then by the theory of faithfully flat descent, y already generates a normal basis of A. (Look at the R-homomorphism $R[C_{p^n}] \to A$, $\sigma \longmapsto \sigma(y)$, and use that it becomes an isomorphism on base extension by S_n.) The point of the argument is hence to find a generator of a normal basis "upstairs" which comes from "downstairs".

A' carries the trivial descent datum $(\Phi_\gamma)_{\gamma \in \Gamma_n}$ (recall $\Gamma_n = \mathrm{Aut}(S_n/R)$), given by $\Phi_\gamma(s \otimes a) = \gamma(s) \otimes a$. The ring $1 \otimes A$ ($\cong A$) is precisely the fixed ring under all Φ_γ (proof of **0**.7.2). It will hence suffice to find a generator z of a normal basis of A'/S_n which is fixed under all Φ_γ, $\gamma \in \Gamma_n$.

On the other hand, since $A' \in NB(R, C_{p^n})$, A' is by **0**.6.5 isomorphic to a Kummer extension $S_n(p^n; u)$ for some unit u of S_n. Recall the corollary to **0**.6.5 which tells us that every y of the form $y = u_0 + u_1\alpha + \ldots + u_{p^n-1}\alpha^{p^n-1}$, with $u_0, \ldots, u_{p^n-1} \in S_n^*$ arbitrary, generates a normal basis of $S_n(p^n; u)$ over S_n (α was short for the p^n-th root of u). By abuse of notation, we consider the Φ_γ also as a descent datum on $S_n(p^n; u)$. It is now our task to find a y of the above shape which is stable under all Φ_γ, and we must do this without knowing exactly just how the Φ_γ act on $S_n(p^n; u) = S_n \oplus S_n\alpha \oplus \ldots \oplus S_n\alpha^{p^n-1}$.

Recall that we have an embedding $\omega: \Gamma_n \to (\mathbb{Z}/p^n)^*$, given by $\gamma(\zeta_n) = \zeta_n^{\omega(\gamma)}$ for $\gamma \in \Gamma_n$. (Here ζ_n is a root of Φ_n in S_n, fixed once and for all).

Lemma 2.2. *Suppose $\gamma \in \Gamma_n$, and φ is any γ-linear R-automorphism of $S_n(p^n;u)$ which commutes with the C_{p^n}-action. Then there exists a unit v of S_n with $\varphi(\alpha) = v\cdot\alpha^{\omega(\gamma)}$.*

Proof. By construction, $S_n\alpha^i$ is precisely the $\zeta_n{}^i$-eigenspace of the generator σ of C_{p^n} on $S_n(p^n;u)$, $0 \le i < p^n$. Now if $\sigma(x) = \zeta_n\cdot x$, then $\sigma(\varphi(x)) = \varphi(\sigma(x)) = \varphi(\zeta_n\cdot x) = \gamma(\zeta_n)\cdot\varphi(x) = \zeta_n{}^{\omega(\gamma)}\cdot\varphi(x)$, hence φ maps the ζ_n-eigenspace of σ into the $\zeta_n{}^{\omega(\gamma)}$-eigenspace of σ. A similar argument for φ^{-1} shows: φ induces an isomorphism of $S_n\alpha$ with $S_n\alpha^{\omega(\gamma)}$. (Note that the latter expression makes sense, even though $\omega(\gamma)$ is only defined mod p^n, since $\alpha^{p^n} = u$ is a unit of S_n.) Hence φ has to map the S_n-generator α of $S_n\alpha$ to some S_n-generator of $S_n\alpha^{\omega(\gamma)}$, which proves the lemma.

Now we are ready for the main part of the proof of Thm. 2.1. Let $I = \{0,1,\ldots,p^n-1\}$. Then I is in canonical bijection with $\mathbb{Z}/p^n$. For $y = \sum_{i\in I} u_i\cdot\alpha^i$, $u_i \in S_n$, we define:

y is *admissible* if all u_i are either units in S_n or zero;

$\mathrm{supp}(y) = \{i \in I \mid u_i \neq 0\}$.

The following remark is obvious: If y and y' are admissible and $\mathrm{supp}(y)$ is disjoint from $\mathrm{supp}(y')$, then $y+y'$ is again admissible, and $\mathrm{supp}(y+y') = \mathrm{supp}(y)\cup\mathrm{supp}(y')$. Now remember that we are looking for an element $y \in S_n(p^n;u)$ which is

a) admissible with $\mathrm{supp}(y) = I$, and

b) stable under all Φ_γ, $\gamma \in \Gamma_n$.

The group Γ_n operates via ω on I (identify I with $\mathbb{Z}/p^n$, and let $(\mathbb{Z}/p^n)^*$ act by multiplication on $\mathbb{Z}/p^n$). In the light of the above obvious remark, it suffices to prove:

(*) If $J \subset I$ is any Γ_n-orbit, then there exists $y_J \in S_n(p^n;u)$, admissible with $\mathrm{supp}(y_J) = J$, and y_J stable under all Φ_γ.

Let $J = \Gamma_n j$ be any orbit, and $H = \mathrm{Stab}(j) \subset \Gamma_n$. By Lemma 2.3 below, there exists $k \in I$ with $j = \sum_{\gamma\in H}\gamma(k)$ $\left(= \sum_{\gamma\in H}\omega(\gamma)\cdot k \text{ by definition}\right)$. Let $z = \prod_{\gamma\in H}\Phi_\gamma(\alpha^k)$. By Lemma 2.2, z has the form

$$z = v\cdot\alpha^{\sum_{\gamma\in H}\omega(\gamma)k} = v'\cdot\alpha^j \quad \text{for units } v,\ v' \text{ of } S_n.$$

Moreover, z is invariant under all Φ_γ, $\gamma \in H$. (Reason: $\Phi_\gamma\Phi_{\gamma'} = \Phi_{\gamma\gamma'}$ for all γ, γ' $\in \Gamma_n$.) Let now $\gamma_1,\ldots,\gamma_t$ be a system of representatives of Γ_n mod H. Then the numbers $\omega(\gamma_i)\cdot j$ $(1 \le i \le t)$ are pairwise distinct and make up the orbit J. By Lemma 2.2 again, $y_J = \Phi_{\gamma_1}(z) + \ldots + \Phi_{\gamma_t}(z)$ is admissible with support J. Moreover by construction, y_J is now stable under all Φ_γ, $\gamma \in \Gamma_n$. This proves (*), and hence the theorem, modulo Lemma 2.3.

Lemma 2.3. *Let $(\mathbb{Z}/p^n)^*$ operate by multiplication on $\mathbb{Z}/p^n$, and let H be any subgroup of $(\mathbb{Z}/p^n)^*$.*

a) If $p \neq 2$, then every $j \in \mathbb{Z}/p^n$ fixed under H has the form $j = \sum_{h\in H} h{\cdot}k$ for some $k \in \mathbb{Z}/p^n$.

b) For $p = 2$, the statement of a) remains correct provided $H \subset \overline{1+4\mathbb{Z}} \subset (\mathbb{Z}/2^n)^*$.

(In more learned terms: The zeroth Tate cohomology group of H with coefficients in $\mathbb{Z}/p^n$ is trivial.)

Proof. If j is incongruent to 0 (mod p^n), then only $h = \bar{1}$ fixes j, so H must be 1, and the lemma is trivial.

Write therefore $j = w{\cdot}p^s$, w prime to p, $1 \le s \le n$. Then H must be contained in $\overline{1+p^{n-s}\mathbb{Z}} \subset \mathbb{Z}/p^n\mathbb{Z}$. For odd p, H is hence necessarily of the form $\overline{1+p^{n-s'}\mathbb{Z}}$ with $s' \le s$, and $|H| = p^{s'}$. (This uses that $(\mathbb{Z}/p^n)^*$ is cyclic.) We get:

$$\sum_{h\in H} h = p^{s'} + \sum_{i=0}^{p^{s'}-1} i{\cdot}p^{n-s'} = p^{s'} + p^{s'}\frac{p^{s'}-1}{2}{\cdot}p^{n-s'} \equiv p^{s'} \pmod{p^n}.$$

Hence, putting $k = w{\cdot}p^{s-s'}$, we get $j = \sum_{h\in H} h{\cdot}k$ as wanted.

For $p = 2$, we have $H \subset \overline{1+4\mathbb{Z}}$ which is cyclic, hence as in the last paragraph H has the form $\overline{1+2^{n-s'}\mathbb{Z}}$. The same calculation as above gives

$$\sum_{h\in H} h \equiv 2^{s'} + 2^{n-1} \pmod{2^n},$$

and the proof can be concluded similarly, using $s' < n-1$.

Remark. What we have proved here for the base extension S_n/R, remains by no means true for arbitrary faithfully flat base extensions. Even if we just replace S_n by S_{n+1} (and retain the Galois group C_{p^n}), the proof is spoiled, and the theorem no longer holds as can be shown.

We obtain a consequence concerning cyclotomic p-extensions which was first proved by Kersten and Michaliček.

Proposition 2.4. *Let $n \ge 1$ and B_n be the n-th cyclotomic p-extension of R. Then B_n has normal basis over R. More concretely, $j_n(B_n)$ is a Kummer extension $S_n(p^n;z)$ with z a primitive p^{n_0}-th root of unity in S_n, i.e. the multiplicative order of z is exactly p^{n_0}.*

Proof. It suffices, by virtue of **0**.6.5 and Thm. 2.1, to prove the second sentence. Now B_n is the fixed ring of the non-p-part of Γ_{n+n_0} in S_{n+n_0}, and (by reasons of rank) $S_1 \subset S_{n+n_0}$ is the fixed ring of the p-primary part of Γ_{n+n_0}, hence by the main theorem: $S_1 \otimes_R B_n \cong S_{n+n_0}$, or in other words (since $S_1 = S_{n_0}$):

$$S_{n_0} \otimes_R B_n \cong S_{n+n_0} \text{ as } C_{p^n}\text{-extensions of } S_{n_0}.$$

We now perform a base extension from $S_1 = S_{n_0}$ to S_n: Let B = tensor product of S_n over S_1 with S_{n+n_0}, or, what is the same, the tensor product of S_n over R with

B_n. Then B is C_{p^n}-Galois over S_n and can hence be written as ι^*B' for some inclusion $\iota: C_{p^s} \subset C_{p^n}$ and some connected $B' \in \mathrm{H}(S_n, C_{p^s})$. Then B' is connected, separable and generated by a root of Φ_{n+n_0} over S_n, hence, by 1,4 and 1.6, B' is the p^{n+n_0}-th cyclotomic extension of S_n as well as of R. By 1.7 b), $p^s = [B':S_n] = p^{n_0}$. By 1.6 c) we therefore have $B' \cong S_n[Z]/(Z^{p^{n_0}} - \zeta)$ for some primitive p^n-th root of unity (i.e. root of Φ_n) $\zeta \in S_n$. We lost track of the operation of the Galois group, but we we can get it back now. To wit, $\overline{Z}$ is a root of Φ_{n+n_0}, hence it generates $\mu_{p^{n+n_0}}(B')$. Let τ be a generator of C_{p^s}. Then $\tau(\overline{Z}) = \overline{Z}^u$ for some u prime to p. Since $z' = Z^{p^{n_0}}$ is in S_n, hence fixed under τ, we must have $u \equiv 1 \pmod{p^n}$. On the other hand, since τ has order p^{n_0}, we cannot have $u \equiv 1 \pmod{p^{n+1}}$, hence $\tau(\overline{Z}) = \eta \cdot \overline{Z}$ with η an element of multiplicative order p^{n_0}. This means: On correct choice of generator τ of C_{p^s} and $\zeta_{n_0} \in S_n$, B' is precisely the Kummer extension $S_n(p^{n_0}; z')$. It is straightforward to show from this that (on correct choice of $\zeta_n \in S_n$) $B = \iota^*B'$ is the Kummer extension $S_n(p^n; z)$ with $z = z'^{p^{n-n_0}} = \overline{Z}^{p^n}$, and z has multiplicative order p^{n_0}. Since $B = j_n(B_n)$, this concludes the proof. (An explanation is due why we are free to choose a generator of C_{p^n} and a primitive p^n-th root of unity in S_n. Such a choice has to be made anyway in the definition of the Kummer extension $S_n(p^n; z)$, and one checks that changes of generator or ζ_n result in replacing z by a power z^u, u prime to p, so nothing is changed.)

Remark. This argument is a little involved, which is not too surprising when one looks at the rather complicated explicit normal bases of S_n/R which are exhibited in [Kersten - Michaliček (1989) §2] or [Greither (1988) II §5].

We now have to consider certain operations on Galois groups. Suppose that T/R is a Γ-Galois extension, for Γ any finite group, and let $n \geq 1$.

Proposition 2.5. *Γ operates canonically by group automorphisms on* $\mathrm{H}(R, C_{p^n})$ *and* $\mathrm{NB}(R, C_{p^n})$.

Proof. For $\gamma \in \Gamma$ and any T-module M we may define

$$M^\gamma = T \otimes_{T,\gamma} M \text{ (the base-changed module along } \gamma\text{: } T \to T).$$

The right-hand module is via β: $1 \otimes x \longmapsto x$ isomorphic to M as an abelian group, and as a "twisted" T-module: more precisely, one lets $s \in T$ act on M via multiplication by $\gamma^{-1}(s)$. To see that with this definition β becomes T-linear, it is enough to note that by definition of the tensor product, $s \otimes x = 1 \otimes \gamma^{-1}(s)x$ in M^γ. If one is a bit confused by this, one may always think of the case that M is an ideal of T; then M^γ is canonically isomorphic to the ideal $\gamma(M) \subset T$.

It is plain that M^γ inherits all structures M might have (T-algebra, C_{p^n}-extension of T), and that $M \longmapsto M^\gamma$ is functorial, so that we do get a map $\mathrm{H}(R, C_{p^n}) \to \mathrm{H}(R, C_{p^n})$ induced by $M \longmapsto M^\gamma$. Furthermore, $(M^\gamma)^\delta$ is canonically isomorphic to $M^{\gamma\delta}$ for γ, $\delta \in \Gamma$ (associativity of $\otimes$, or an easy direct argument), so we get an operation of Γ on $\mathrm{H}(R, C_{p^n})$. This is an operation by group automorphisms, and preserves normal bases, since any base change preserves Harrison product and normal bases (**0**.3.11 and **0**.6.4).

The group Γ operates also on several other objects associated with T: first of all, on T^*, and on T^*/m (which was defined as T^*/T^{*m}) for any $m \in \mathbb{N}$. Second, Γ operates on the group $\mathrm{Disc}(T;m)$ of discriminant modules as follows: for (P,μ) in $\mathrm{Disc}(T;m)$ and $\gamma \in \Gamma$, let $(P,\mu)^\gamma = (P^\gamma, \alpha\mu^\gamma)$, where $(P^\gamma)^{\otimes m}$ has to be identified with $(P^{\otimes m})^\gamma$, μ^γ is the map $(P^{\otimes m})^\gamma \longrightarrow T^\gamma$ induced by μ, and α: $T^\gamma = T \otimes_\gamma T \to T$ is the canonical isomorphism $s \otimes t \longmapsto s\gamma(t)$. (If one uses the definition "P^γ is just P, with T-multiplication twisted by γ^{-1}" as above, then μ^γ is just μ, and one may omit α. But this is slightly dangerous, since now the range of μ^γ is T^γ, i.e. T with twisted T-multiplication, an object likely to cause confusion.)

Recall from the proof of **0**.5.4 the canonical map i': $T^*/m \longrightarrow \mathrm{Disc}(T;m)$ defined by $i'[u] = (T, \tilde{u})$ where $\tilde{u}$ stands for the map $T^{\otimes m} = T \longrightarrow T$.

Lemma 2.6. *The homomorphism i': $T^*/m \longrightarrow \mathrm{Disc}(T;m)$ is Γ-equivariant.*

Proof. Let $u \in T^*$, and $\gamma \in \Gamma$. We calculate $i'[u]^\gamma$. We have $i'[u]^\gamma = (T\otimes_\gamma T, T\otimes_\gamma \tilde{u})$, and we have the T-isomorphism β: $T\otimes_\gamma T \longrightarrow T$, $s\otimes t \longmapsto s\gamma(t)$. We claim that α induces an isomorphism of discriminant modules $(T\otimes_\gamma T, T\otimes_\gamma \tilde{u}) \to (T, \gamma(\tilde{u}))$. There is a diagram

$$\begin{array}{ccccccc}
T\otimes_\gamma T = T\otimes_\gamma (T^{\otimes m}) & = & (T\otimes_\gamma T)^{\otimes m} & \xrightarrow{\alpha^{\otimes m}} & T^{\otimes m} & = & T \\
\big\downarrow{\scriptstyle T\otimes\tilde{u}} & & & & & & \big\downarrow{\scriptstyle \gamma(\tilde{u})} \\
T\otimes_\gamma T & & & & & & \\
\big\downarrow{\scriptstyle \alpha} & & & & & & \\
T & & & = & & & T
\end{array}$$

of T-linear maps; if we start in the upper left hand corner and chase $1\otimes 1$ both ways, we obtain $\gamma(u)$ both times, so the diagram commutes, and the claim follows. Hence $i'[u]^\gamma = $ class of $(T, \gamma(\tilde{u})) = i'(\gamma(u))$, q.e.d.

Now we return to the Galois extension S_n/R with group Γ_n. We constructed an embedding ω: $\Gamma_n \to (\mathbb{Z}/p^n)^*$. We now define a twisting operation for certain Γ_n-modules.

Definition. Let M be any Γ_n-module with $p^n M = 0$. One constructs a new Γ_n-module $M(-1)$ by the following prescription:

$M(-1) \;=\; M$ as an abelian group;

the action $*$: $\Gamma \times M(-1) \to M(-1)$ is given by $\gamma * x = \omega(\gamma)^{-1} \cdot \gamma(x)$ $\quad (\gamma \in \Gamma_n, x \in M)$.

Remarks. a) The condition $p^n M = 0$ ensures that multiplication with $\omega(\gamma)^{-1} \in \mathbb{Z}/p^n$ makes sense. It is then obvious that $M(-1)$ is again a Γ_n-module.

b) There is an obvious generalization for all $k \in \mathbb{Z}$: $M(k) = M$ as an abelian group, with Γ_n-action $\gamma * x = \omega(\gamma)^k \cdot \gamma(x)$. For $k > 0$, there is the following more conceptual way of viewing $M(k)$:

$$M(k) \cong M \otimes_{\mathbb{Z}} \mathbb{Z}/p^n \otimes_{\mathbb{Z}} \ldots \otimes_{\mathbb{Z}} \mathbb{Z}/p^n \quad (k \text{ factors } \mathbb{Z}/p^n),$$

where Γ_n operates diagonally, and via ω on each factor $\mathbb{Z}/p^n$. One may just as well write $\mu_{p^n}(S_n)$ in the place of $\mathbb{Z}/p^n$: this gives automatically the correct Γ_n-action.

c) This whole construction is well-known in number theory. $M(k)$ is usually called the k-fold Tate twist of M.

The first appearance of this twist is in the Kummer sequence. Note that S_n is p^n-kummerian. By **0**.5.4 we have a short exact sequence of abelian groups (note that we have introduced two twists, but this does not affect the exactness, as long as we view the sequence just as a sequence of abelian groups):

$$1 \longrightarrow \left(S_n^*/p^n\right)(-1) \xrightarrow{\;i\;} \mathrm{H}(S_n, C_{p^n}) \xrightarrow{\;\pi\;} \mathrm{Pic}(S_n)[p^n](-1) \longrightarrow 1.$$

Proposition 2.7. *This sequence is also a sequence of Γ_n-modules, i.e. i and π are Γ_n-equivariant.*

Proof. By **0**.5.3, there is an isomorphism d: $\mathrm{Disc}(S_n; p^n) \longrightarrow \mathrm{H}(S_n, C_{p^n})$ and a commutative diagram

$$\begin{array}{ccccccccc} & & & \overset{i}{\nearrow} & \mathrm{H}(S_n, C_{p^n}) & \overset{\pi}{\searrow} & & & \\ 1 & \longrightarrow & S_n^*/p^n & & \cong\; d & & \mathrm{Pic}(S_n)[p^n] & \longrightarrow & 1. \\ & & & \underset{i'}{\searrow} & \mathrm{Disc}(S_n; p^n) & \underset{\pi'}{\nearrow} & & & \end{array}$$

We know already (Lemma 2.6) that i' is Γ_n-equivariant. It is clear from the definitions that π' is Γ_n-equivariant ($\pi'(P,\mu) = P$.) Hence it will suffice to establish that d: $\mathrm{Disc}(S_n; p^n)(-1) \longrightarrow \mathrm{H}(S_n, C_{p^n})$ is Γ_n-equivariant.

Let $(P,\mu) \in \mathrm{Disc}(S_n; p^n)$ and $A = d(P,\mu)$. In the notation of **0** §5:

$$\begin{aligned} d(P,\mu) &= S_n(p^n; P,\mu) \\ &= S_n \oplus P \oplus \ldots \oplus P^{\otimes(p^n-1)} \text{ (multiplication induced by } \mu). \end{aligned}$$

The fixed generator σ of C_{p^n} acts as multiplication by ζ_n on P, where ζ_n is a fixed generator of $\mu_{p^n}(S_n)$. We apply $\gamma \in \Gamma_n$ and obtain:

$$\begin{aligned} d(P,\mu)^\gamma &= S_n^\gamma \oplus P^\gamma \oplus \ldots \oplus (P^{\otimes(p^n-1)})^\gamma \text{ (mult. induced by } \mu^\gamma) \\ &= S_n \oplus P^\gamma \oplus \ldots \oplus (P^{\otimes(p^n-1)})^\gamma \text{ (mult. induced by } \alpha\mu^\gamma), \end{aligned}$$

and it would be tempting to infer that $d(P,\mu)^{\gamma}$ is just $d((P,\mu)^{\gamma})$. But this is not true as we shall see presently. For let $f\colon A = S_n(p^n; P,\mu) \longrightarrow A$ be the action of $\sigma \in \Gamma_n$. Then the action of σ on $A^{\gamma} = S_n(p^n; P,\mu)^{\gamma}$ is by $1\otimes_{\gamma} f$, i.e. the elements of $P^{\gamma} = S_n \otimes_{\gamma} P$ are multiplied by $1\otimes\zeta_n$ under σ, and by definition of the tensor product, $1\otimes\zeta_n = \zeta_n^{\omega(\gamma)}\otimes 1$. Now the S_n-module structure on A^{γ} comes from the tensor factor S_n, thus σ acts on P^{γ} via multiplication by $\zeta_n^{\omega(\gamma)}$, and *not* by ζ_n as on $d((P,\mu)^{\gamma}) = S_n(p^n;(P,\mu)^{\gamma})$.

Let $t \in \{0,\dots,p^n-1\}$ with $\bar{t} = \omega(\gamma)^{-1}$. We have to show $d(t\cdot(P,\mu)^{\gamma}) = d(P,\mu)^{\gamma}$, i.e. the extension $B = S_n \oplus (P^{\gamma})^{\otimes t} \oplus \dots \oplus (P^{\gamma})^{\otimes t(p^n-1)}$ (multiplication induced by $\mu^{\otimes t}$, σ acts as ζ_n on $(P^{\gamma})^{\otimes t}$) is isomorphic to A^{γ} in $\mathrm{H}(S_n, C_{p^n})$. The identity map $(P^{\gamma})^{\otimes t} \to (P^{\gamma})^{\otimes t}$ induces a S_n-algebra homomorphism $f\colon B \to A^{\gamma}$. Now in A^{γ}, σ acts as $\zeta_n^{\omega(\gamma)}$ on P^{γ}, hence it acts as $\zeta_n^{\omega(\gamma)\cdot t} = \zeta_n$ on $(P^{\gamma})^{\otimes t} \subset A^{\gamma}$, hence f is also C_{p^n}-equivariant, and by **0**.1.12 we finally get $B \cong A^{\gamma}$, q.e.d.

Remark. It is equally possible to show the Γ_n-equivariance of i and π directly, without using d.

Combining the last result with **0**.6.5. yields:

Corollary 2.8 *The map i induces a Γ_n-isomorphism*

$$(S_n/p^n)(-1) \xrightarrow{\cong} \mathrm{NB}(S_n, C_{p^n});$$

the map π induces a Γ_n-isomorphism

$$\mathrm{Pic}(S_n)[p^n](-1) \xrightarrow{\cong} \mathrm{P}(S_n, C_{p^n}).$$

We need a few easy preparations for the descent back to R which will be studied in the next section.

Lemma 2.9. *Let R, S_n, Γ_n be as before, assume $p \neq 2$, and let β be a generator of Γ_n. Then there exists a lifting $w \in \mathbb{Z}$ of $\omega(\beta) \in (\mathbb{Z}/p^n)^*$ such that with $m = |\Gamma_n|$ one has*

$$w^m \equiv 1 \bmod p^n, \text{ but } w^m \not\equiv 1 \bmod p^{n+1}.$$

Proof. (Well-known.) Let $s = \varphi(p^n)/m \in \mathbb{N}$. By the theory of cyclic groups, $\omega(\beta)$ can be written in the form c^s, c a generator of $(\mathbb{Z}/p^n)^*$. Now c lifts to a generator c' of $(\mathbb{Z}/p^{n+1})^*$. Write $c' = \overline{w_0}$, $w_0 \in \mathbb{N}$. Let $w = w_0^s$. Then w is a lifting of $\omega(\beta)$, and certainly $w^m \equiv 1 \bmod p^n$. But $w^m = w_0^{\varphi(p^n)}$, so $w^m \bmod p^{n+1}$ equals $c'^{\varphi(p^n)}$ which is not 1 since $\varphi(p^n) < \varphi(p^{n+1}) = \mathrm{ord}\,(\mathbb{Z}/p^{n+1})^*$.

Definition and Lemma 2.10. *Let β be a generator of Γ_n, $m = |\Gamma_n|$, and w a lifting of $\omega(\beta)$ chosen as in Lemma 2.8. Define*

$$\xi = \sum_{i=0}^{m-1} w^i \cdot \beta^{-i} \in \mathbb{Z}[\Gamma_n].$$

Then $(1 - w\beta^{-1})\cdot\xi = 1 - w^m$, and $1 - w^m$ is divisible by p^n but not by p^{n+1}.

Proof. $(1 - w\beta^{-1})\cdot\xi = 1 - w^m\beta^{-m} = 1 - w^m$. The rest follows from Lemma 2.9.

§3 Cyclotomic descent: the main results

In this section we fix a prime $p \neq 2$. Let $n \geq 1$.

Let R be connected with $p^{-1} \in R$, and let S_n the p^n-th cyclotomic extension of R as in §2. Recall that Γ_n is the Galois group of S_n over R. We begin with the following observation:

Lemma 3.1. *The image of the base extension map j: $\mathrm{H}(R, C_{p^n}) \longrightarrow \mathrm{H}(S_n, C_{p^n})$ is contained in $\mathrm{H}(S_n, C_{p^n})^{\Gamma_n}$ (= fixed subgroup of Γ_n). Similarly, the image of the map j: $\mathrm{NB}(R, C_{p^n}) \longrightarrow \mathrm{NB}(S_n, C_{p^n})$ is contained in $\mathrm{NB}(S_n, C_{p^n})^{\Gamma_n}$.*

Proof. If $\gamma \in \Gamma$ and $A \in \mathrm{H}(R, C_{p^n})$, then we have to show that $j(A) = S_n \otimes_R A$ is fixed under γ (up to isomorphism). This is easily seen as follows:

$$(S_n \otimes_R A)^{\gamma} = S_n \otimes_{\gamma} S_n \otimes_R A \cong S_n \otimes_R A,$$

since the composition $R \rightarrow S_n \overset{\gamma}{\rightarrow} S_n$ is the same as the inclusion $R \rightarrow S_n$. The statement concerning NB is proved in exactly the same way.

It is now an interesting question how far $\mathrm{Im}(j)$ is away from $\mathrm{H}(S_n, C_{p^n})^{\Gamma_n}$. Let from now on j denote the base extension map $\mathrm{H}(R, C_{p^n}) \rightarrow \mathrm{H}(S_n, C_{p^n})^{\Gamma_n}$ and j_0: $\mathrm{NB}(R, C_{p^n}) \rightarrow \mathrm{NB}(S_n, C_{p^n})^{\Gamma_n}$ its restriction. We also have an induced map $\bar{j}$: $\mathrm{P}(R, C_{p^n}) \longrightarrow \mathrm{P}(S_n, C_{p^n})^{\Gamma_n}$. There is the following commutative diagram:

$$\begin{array}{ccccccccc}
 & & 1 & & 1 & & & & \\
 & & \downarrow & & \downarrow & & & & \\
 & & \mathrm{Ker}(j_0) & \longrightarrow & \mathrm{Ker}(j) & & & & \\
 & & \downarrow & & \downarrow & & & & \\
1 & \longrightarrow & \mathrm{NB}(R, C_{p^n}) & \longrightarrow & \mathrm{H}(R, C_{p^n}) & \longrightarrow & \mathrm{P}(R, C_{p^n}) & \longrightarrow & 1 \\
 & & {\scriptstyle j_0}\downarrow & & {\scriptstyle j}\downarrow & & {\scriptstyle \bar{j}}\downarrow & & \\
1 & \longrightarrow & \mathrm{NB}(S_n, C_{p^n})^{\Gamma_n} & \longrightarrow & \mathrm{H}(S_n, C_{p^n})^{\Gamma_n} & \longrightarrow & \mathrm{P}(S_n, C_{p^n})^{\Gamma_n} & & \\
 & & \cong\downarrow & & \cong\downarrow & & \cong\downarrow & & \\
1 & \longrightarrow & (S_n^*/p^n)(-1)^{\Gamma_n} & \rightarrow & \mathrm{Disc}(S_n, p^n)(-1)^{\Gamma_n} & \rightarrow & \mathrm{Pic}(S_n)[p^n](-1)^{\Gamma_n} & & \\
 & & \downarrow & & \downarrow & & & & \\
 & & \mathrm{Coker}(j_0) & \longrightarrow & \mathrm{Coker}(j) & & & & \\
 & & \downarrow & & \downarrow & & & & \\
 & & 1 & & 1\,. & & & &
\end{array}$$

Note that there is no 1 at the right of the fourth and fifth line, since taking Γ_n-invariants is only a left exact functor. We first compare j_0 and j:

Lemma 3.2. *a)* $\mathrm{Ker}(j_0) \cong \mathrm{Ker}(j)$ *via the map in the diagram.*

b) The map $\mathrm{Coker}(j_0) \longrightarrow \mathrm{Coker}(j)$ *in the diagram is injective.*

Proof. a) The map $\mathrm{Ker}(j_0) \to \mathrm{Ker}(j)$ is trivially injective. If $A \in \mathrm{Ker}(j)$, then $j(A)$ is trivial, hence lies in $\mathrm{NB}(S_n, C_{p^n})$. By Thm. 2.1, A is in $\mathrm{NB}(R, C_{p^n})$, whence indeed $A \in \mathrm{Ker}(j_0)$.

b) Suppose $A \in \mathrm{NB}(S_n, C_{p^n})^{\Gamma_n}$ such that $A \in \mathrm{Im}(j)$. Since any preimage under j of A has a normal basis by 2.1, we have already $A \in \mathrm{Im}(j_0)$, q.e.d.

The main results of this section and chapter are, besides 3.7 below, the following two theorems:

Theorem 3.3. $\mathrm{Ker}(j)$ *and* $\mathrm{Coker}(j)$ *are both cyclic of order* $m' = \gcd(p^n, |\Gamma_n|)$.

Theorem 3.4. $\mathrm{Ker}(j_0)$ *and* $\mathrm{Coker}(j_0)$ *are both cyclic of order* $m' = \gcd(p^n, |\Gamma_n|)$. *With* $\xi' = \sum_{\gamma \in \Gamma_n} \omega(\gamma)\cdot\gamma^{-1} \in (\mathbb{Z}/p^n)[\Gamma_n]$ *one has:*

$\mathrm{Im}(j_0)$ *corresponds to* $\xi'\cdot(S_n^*/p^n)$ *via* $\mathrm{NB}(S_n, C_{p^n}) \cong S_n^*/p^n$. (Prop. **0** 6.5)

(The group ring $(\mathbb{Z}/p^n)[\Gamma_n]$ *acts on the multiplicative group* S_n^*/p^n *as follows: for* $\eta = \sum_\gamma a_\gamma\cdot\gamma \in (\mathbb{Z}/p^n)[\Gamma_n]$, $x \in S_n^*/p^n$, *one defines* $\eta\cdot x = \prod_\gamma \gamma(x)^{a_\gamma}$.*)*

Remark. For the case $R = K$ a field, 3.4 goes back to Miki (1974), Saltman (1982), and Childs (1984). In this case $\mathrm{NB}(K, C_{p^n}) = \mathrm{H}(R, C_{p^n})$ since the Picard group of the artinian ring $K[C_{p^n}]$ is zero (or by the normal basis theorem in classical Galois theory), so 3.3 is covered by 3.4 in the field case.

The proofs of 3.3 and 3.4 will be quite different in nature. For 3.3 we shall use a quick argument in the style of Galois cohomology. For 3.4 we give a very explicit calculation which seems of interest in itself. It is not clear at the moment how one could, maybe, deduce 3.4 directly from 3.3; what is lacking here is an easy proof that $\mathrm{Coker}(j_0) \longrightarrow \mathrm{Coker}(j)$ is onto.

Proof of **3.3.** We use the existence of a separable closure R^{sep} of R; recall $\Omega = \Psi_R = \mathrm{Aut}(R^{sep}/R)$, and the isomorphism of functors on finite abelian groups $\mathrm{H}(R,-) \cong \mathrm{Hom}_{cont}(\Omega,-)$ from **0**§3. The extension S_n of R is embedded in R^{sep}; let $\Omega_n \subset \Omega$ be the fixed group of S_n. Let $C = C_{p^n}$. We claim that j: $\mathrm{H}(R,C) \longrightarrow \mathrm{H}(S_n,C)$ corresponds to the restriction *res*: $\mathrm{Hom}_{cont}(\Omega, C) \longrightarrow \mathrm{Hom}_{cont}(\Omega_n, C)$.

Proof of this claim: Note first that R^{sep} is automatically a separable closure of S_n (by transitivity of separability), and $\Omega_n = \mathrm{Aut}(R^{sep}/S_n)$, so that we do have a canonical isomorphism $\mathrm{H}(S_n,-) \cong \mathrm{Hom}_{cont}(\Omega_n,-)$. Let f: $\Omega \to C$ be continuous, and $g = f|\Omega_n$. As shown in **0**.3.7, f corresponds to the Galois extension f_E^*E, where E is a Galois extension of R inside R^{sep} such that f factors over f_E: $\mathrm{Aut}(E/R) \to C$. We may choose an E which contains S_n. Then g_E is just the restriction of f_E to $\mathrm{Aut}(E/S_n) \subset \mathrm{Aut}(E/R)$, and $g_E^{\,*}E$ is a C-Galois extension of S_n. Restriction gives an R-algebra map $f_E^{\,*}E \longrightarrow g_E^{\,*}E$; we hence obtain a C-equivariant S_n-algebra homomorphism α: $S_n \otimes_R f_E^{\,*}E \longrightarrow g_E^{\,*}E$ which has to be an isomorphism by **0**.1.12. Hence g corresponds to $S_n \otimes_R f_E^{\,*}E = j(f_E^{\,*}E)$, which proves the claim.

By Galois theory, $\Omega/\Omega_n = \Gamma_n$, the automorphism group of S_n/R. It is now easy to transfer the Γ_n-action on $\mathrm{H}(S_n,C)$ to $\mathrm{Hom}_{cont}(\Omega_n,C)$: If B/S_n is connected and C-Galois, without loss $B \subset R^{sep}$, then one sees easily that for $\gamma \in \Gamma_n$, $B^\gamma = \gamma'(B)$, where γ' is any lift of γ to Ω. Now B belongs to some continuous f: $\Omega_n \to C$, and for each $\sigma \in \Omega_n$, $f(\sigma)$ is the (!) element of C which acts like σ on B. From this description we get that B^γ belongs to ${}^{\gamma'}\!f = (\sigma \longmapsto \gamma' f(\gamma'^{-1}\sigma))$. If B is nonconnected, then B comes from $\mathrm{H}(S_n,C')$, $C' \subset C$ a proper subgroup, and by induction on $|C|$, the last formula is still correct. Hence Γ_n operates on $\mathrm{Hom}_{cont}(\Omega_n,C)$ by the prescription $(\gamma,f) \longmapsto {}^{\gamma'}\!f$, and this is the usual operation of $\Gamma_n = \Omega/\Omega_n$ on $\mathrm{Hom}_{cont}(\Omega_n,C)$.

We now have to use a bit of group cohomology (cf. for example Serre (1964) Chap. I or Neukirch (1969)). All modules used as right arguments of H^1 and H^2 will be trivial modules over the profinite group which is the left argument. Let inf_i $(i = 1,2)$ be the inflation maps $\mathrm{H}^i(\Omega/\Omega_n,C) \longrightarrow \mathrm{H}^i(\Omega,C)$, and recall that for any profinite group Δ and any *trivial* Δ-module D, $\mathrm{H}^1(\Delta,D) = \mathrm{Hom}_{cont}(\Delta,D)$. It is well-known that the image of the restriction *res*: $\mathrm{H}^1(\Omega,C) \longrightarrow \mathrm{H}^1(\Omega_n,C)$ is contained in the fixed subgroup of Γ_n acting on $\mathrm{H}^1(\Omega_n,C)$. Using $\Omega/\Omega_n = \Gamma_n$, we obtain the following commutative diagram:

$$
\begin{array}{ccccccccc}
 & & \mathrm{H}(R,C) & \xrightarrow{\;j\;} & \mathrm{H}(S_n,C)^{\Gamma_n} & & & & \\
 & & \cong\big\downarrow & & \cong\big\downarrow & & & & \\
1 \to \mathrm{H}^1(\Gamma_n,C) & \xrightarrow{inf_1} & \mathrm{H}^1(\Omega,C) & \xrightarrow{res} & \mathrm{H}^1(\Omega_n,C)^{\Gamma_n} & \xrightarrow{\;t\;} & \mathrm{H}^2(\Gamma_n,C) & \xrightarrow{inf_2} & \mathrm{H}^2(\Omega,C),
\end{array}
$$

in which the bottom line still has to be explained. Either it is obtained from the low-degree terms in the Hochschild-Serre spectral sequence, or it is established by a direct argument (see e.g. MacLane (1963) [XI 10.6]). The map t is called *transgression*. For the reader's convenience, we give the definition in our case: Identify $\mathrm{H}^1(-,C)$ and $\mathrm{Hom}_{cont}(-,C)$ as above and pick a left transversal $\{\beta_\tau \mid \tau \in \Gamma_n\}$ of Ω mod Ω_n. Define for $f \in \mathrm{Hom}_{cont}(\Omega_n,C)$ and $\tau,\vartheta \in \Gamma_n$:

$$t(f)_{\tau,\vartheta} = f(\beta_\tau \beta_\vartheta \beta_{\tau\vartheta}^{-1}) \in C.$$

Whenever $f \in \mathrm{Hom}_{cont}(\Omega_n,C)^{\Gamma_n}$, $(t(f)_{\tau,\vartheta})$ is a 2-cocycle; its class in $\mathrm{H}^2(\Gamma_n,C)$ is $t(f)$. This cocycle is a coboundary iff f comes by restriction from $\mathrm{H}^1(\Omega,C)$.

Claim: inf_2 is the zero map.

Proof. First case: Γ_n has order prime to p. Then already $H^2(\Gamma_n, C) = 0$ since the orders of Γ_n and C are coprime.

Second case: p divides the order of Γ_n. Then $n \geq 2$, and S_n is larger than S_1, since $[S_1:R] = |\Gamma_1|$ divides $p-1$. Hence the number n_0 (see §1) is not ∞, and $|\Gamma_N| = p^{N-n_0}$ for m large. Hence there exists $N > n$ such that p^n divides $[S_m:S_n]$. Pick such an N. We shall show that already inf': $H^2(\Gamma_N, C) \longrightarrow H^2(\Gamma_n, C)$ is zero.

For any finite cyclic group which acts trivially on C, we have $H^2(G, C) \cong \mathrm{Ext}^1_{\mathbf{Z}}(G, C)$ canonically, and inf' corresponds to the natural map (pulling back of extensions along $\Gamma_N \to \Gamma_n$) from $\mathrm{Ext}(\Gamma_N, C)$ to $\mathrm{Ext}(\Gamma_n, C)$. Starting with any extension $0 \to C \to E \to \Gamma_n \to 0$, let $0 \to C \to E' \to \Gamma_N \to 0$ be the pulled-back extension; we have to show that it splits. Take a preimage e' of a preimage of a generator γ' of Γ_N, and let e be the image of e' in E. Then $e^{|\Gamma_n|}$ lies in C; call it c. We now get $e'^{|\Gamma_N|} = c^{|\Gamma_N|/|\Gamma_n|} \in C$, and the exponent is divisible by $p^n = |C|$, hence $e'^{|\Gamma_N|} = 0$, and $\gamma' \to e'$ gives a splitting of the pulled-back extension, so the claim holds.

From the exactness of the diagram, and the claim, we now obtain

$$\mathrm{Ker}(j) \cong H^1(\Gamma_n, C); \qquad \mathrm{Coker}(j) \cong H^2(\Gamma_n, C).$$

The H^1 term is $\mathrm{Hom}(\Gamma_n, C)$, hence cyclic of order $m = \text{g.c.d.}(|\Gamma_n|, p^n)$. The H^2 term is also cyclic since C and Γ_n are cyclic. Its order is the same as the order of the H^1 term. This is elementary; in learned language this is expressed by saying that the Herbrandt quotient of C is 1. Cf. Neukirch (1964). This concludes the proof of 3.3.

Proof of **Thm. 3.4.** The kernel of j_0 equals the kernel of j by 3.1. By **0**.3.12, $\mathrm{Ker}(j) \cong \mathrm{Hom}(\Gamma_n, C)$, and this easily gives the claimed statement about $\mathrm{Ker}(j)$. (Note that we have reproved a part of 3.3.) The difficult part is determining $\mathrm{Coker}(j_0)$.

Let us say that an element $[x] \in S_n^*/p^n$ *descends*, if $i(x) = S_n(p^n;x)$ (which is an element of $\mathrm{NB}(S_n, C_{p^n})$) comes via j_0 form an element of $\mathrm{NB}(R, C_{p^n})$. As we know, it is the same to ask whether $i(x)$ comes from $H(R, C_{p^n})$ via j.

We first prove: $[x]$ descends iff $[x]$ has the form $[y]^{\xi'}$ for some $y \in S_n^*$ (we switch to exponential notation which is more natural). Let $A = S_n(p^n;x) = \bigoplus_{i=0}^{p^n-1} S_n \cdot z^i$, with $z^{p^n} = x$.

Let $\Gamma = \Gamma_n$ and suppose that $(\Phi_\gamma)_{\gamma\in\Gamma}$ is a descent datum on A. Let β be a generator of Γ and $\varphi = \Phi_\beta$. Then, with $m = |\Gamma|$, the family $(\Phi_\gamma)_{\gamma\in\Gamma}$ is just the family $(\varphi^i)_{0\leq i<m}$. Hence φ must satisfy the following two properties:

(*i*) φ is β-linear, and

(*ii*) φ^m is the identity.

Conversely, if φ is any R-automorphism of A satisfying (i) and (ii), then one obtains a descent datum by letting $\Phi_\gamma = \varphi^i$ for $\gamma = \beta^i$, $i \in Z$. Therefore, $[x]$ descends iff A admits a C_{p^n}-invariant automorphism φ satisfying (i) and (ii).

Any C_{p^n}-invariant φ with (i) satisfies

$$\varphi(z) = v{\cdot}z^w, \quad \text{with } w \in Z \text{ a lifting of } \omega(\beta)$$

by Lemma 2.2. Inserting this formula in itself yields:

$$\varphi^2(z) = \beta(v){\cdot}v^w{\cdot}z^{w^2} = v^{\beta+w}{\cdot}z^{w^2},$$

$$\varphi^3(z) \quad = \quad v^{\beta^2+w\beta+w^2}{\cdot}z^{w^3}, \ldots,$$

and finally for the exponent m:

$$\varphi^m(z) \;=\; v^\eta{\cdot}z^{w^m}, \text{ with } \eta = \beta^{m-1}+w\beta^{m-2}+\ldots+w^{m-1} \in Z[\Gamma].$$

Note that mod p^n, η becomes $\beta^{-1}{\cdot}\xi'$ (since $\beta^m = 1$). If φ satisfies also (ii), we may rewrite the last formula as

$$z^{1-w^m} \;=\; v^\eta.$$

Now we assume that β and w haven been choosen in such a way that $1-w^m = p^nq$ with q prime to p (Lemma 2.9). Since $x = z^{p^n}$, we get the following in S_n^*/p^n:

$$[x]^q \;=\; [v]^\eta = [v]^{\beta^{-1}\xi'},$$

hence we have proved: If x descends, then $[x] = [y]^{\xi'}$ with $[y] = [v]^{q^{-1}\beta^{-1}}$ (note that the exponent q^{-1} makes sense because we are working in S^*/p^n).

Suppose now that, conversely, we have $y \in S_n^*$ with $[x] = [y]^{\xi'}$. It is to be shown that $[x]$ descends. We obtain an equation in S_n^*: $x = u^\eta{\cdot}r^{p^n}$ for some u, $r \in S_n^*$. Since $[x] = [x{\cdot}r^{-p^n}]$, we may assume $r = 1$. With $v = u^q$ we obtain $x^q = v^\eta$. As obove, $x^q = z^{1-w^m}$.

We now *define* an R-algebra homomorphism φ: $A \to A$ by $\varphi(z) = v{\cdot}z^w$ and $\varphi|S_n = \beta$. One has to check that φ is well-defined: the point is to show that $(v{\cdot}z^w)^{p^n} = \beta(x)$. We calculate:

$$\begin{aligned}(v{\cdot}z^w)^{p^n} &= u^{p^nq}{\cdot}z^{p^nw} = u^{p^nq}{\cdot}x^w\\ &= u^{1-w^m}{\cdot}x^w\\ &= u^{\eta(\beta-w)}{\cdot}x^w \quad \text{(since } (\beta-w){\cdot}\eta = \beta^m-w^m = 1-w^m\text{)}\\ &= x^{\beta-w}{\cdot}x^w \qquad \text{(since } u^\eta = x\text{)}\\ &= x^\beta.\end{aligned}$$

We also need that φ satisfies (i) and (ii). The former holds by construction, and the latter property amounts to the formula $\varphi^m(z) = z$. But by the above calculation, $\varphi^m(z) = v^\eta{\cdot}z^{w^m} = v^\eta{\cdot}z{\cdot}z^{-p^nq} = z{\cdot}v^\eta{\cdot}x^{-q}$, and we have $x^q = v^\eta$ by hypothesis. This finishes the first (and main) step, and proves the last sentence in Thm. 3.3.

From the diagram at the beginning of the §, we now have that $E = \xi'(S_n^*/p^n)$ is contained in $E' = (S_n^*/p^n)(-1)^{\Gamma_n}$. We have to show: E'/E is cyclic of order g.c.d.(m,p^n). Assume again β and w are chosen as in 2.9 and the first part of the proof. Since the action of β on $(S_n^*/p^n)(-1)$ is given by $(\beta,[x]) = [\beta(x)^{w^{-1}}]$, the fixed group E' is precisely the annihilator of $1-w^{-1}\beta$, or (equivalently) of $1-w\cdot\beta^{-1}$ on S_n^*/p^n. On the other hand, E is precisely the image of multiplication by $\xi = \sum_{0\le i<m} w^i\beta^{-i}$ on S_n^*/p^n.

By Lemma 2.9, $(1-w\beta^{-1})\xi = p^n q$ with $q \in \mathbb{Z}$ prime to p. This first of all reproves the inclusion $E \subset E'$. We shall define a homomorphism

$$\delta\colon E' \longrightarrow \mu_{p^n}/m$$

(where μ_{p^n} is short for $\mu_{p^n}(S_n)$, and μ_{p^n}/m means of course μ_{p^n} modulo m-th powers), and we then show δ is onto with kernel E, which will suffice.

For $[x] \in E'$, we have $x^{1-w\beta^{-1}} = y^{p^n}$ for some $y \in S_n^*$, hence $x^{q(1-w\beta^{-1})} = y^{qp^n} = y^{\xi(1-w\beta^{-1})}$. Therefore

$$(x^q/y^\xi)^{(1-w\beta^{-1})} = 1.$$

This implies $1 = (x^q/y^\xi)^{(1-w\beta^{-1})} = (x^q/y^\xi)^{qp^n}$, hence $x^{q^2}/y^{q\xi}$ is a p^n-th root of unity. Define

$$\delta[x] = \text{class of } x^{q^2}/y^{q\xi} \text{ in } \mu_{p^n}/m.$$

<u>δ is well-defined:</u> If we take y' in the place of y above, then $y' = \zeta y$ with some $\zeta \in \mu_{p^n}$, and the "new" value of $\delta[x]$ is $\zeta^{-q\xi}$ times the old one. Now each $w^i\beta^{-i}$ is identity on ζ, by definition of w and ω. Hence $\zeta^{-q\xi} = \zeta^{-qm}$, and this factor disappears in μ_{p^n}/m. It is a trivial consequence that δ is a homomorphism.

<u>δ is onto:</u> For $x = \zeta_n$ (a generator of μ_{p^n}), choose $y = 1$ and obtain, going through the definition: $\delta[x] = $ class of $\zeta_n^{q^2}$, which is, of course, a generator of μ_{p^n}/m.

<u>$\mathrm{Ker}(\delta) = E$:</u> If $x = v^\xi$, then $x^{(1-w\beta^{-1})} = v^{p^n q}$, and we may take $y = v^q$ in the definition of $\delta[x]$. Then $x^q/y^\xi = v^{\xi q}/v^{q\xi} = 1$, and $\delta[x] = 1$. If conversely $\delta[x]$ is trivial, we have $x^{q^2} \equiv y^{q\xi}$ modulo a factor ϑ in $\mu_{p^n}{}^m$. The same calculation as in the proof of well-definedness shows: We can change y by a power of ζ_n so as to make disappear ϑ. Hence we may assume $x^{q^2} = y^{q\xi}$, hence $[x]^{q^2} \in E$, which implies $[x] \in E$. Thm. 3.4 is now proved.

Corollary 3.5. *The map* $\mathrm{Coker}(j_0) \longrightarrow \mathrm{Coker}(j)$ *in the diagram preceding* 3.2 *is an isomorphism. (It is monic by* 3.2, *and* $|\mathrm{Coker}(j_0)| = |\mathrm{Coker}(j)|$ *by* 3.3 *and* 3.4.)

We now are interested in the map $\bar{j}$ in the big diagram preceding 3.2. Let us redraw the diagram, simplifying it a little:

$$\begin{array}{ccccccccc}
 & & \mathrm{Ker}(j_0) & \longrightarrow & \mathrm{Ker}(j) & & & & \\
 & & \cap & & \cap & & & & \\
1 & \longrightarrow & \mathrm{NB}(R,C_{p^n}) & \longrightarrow & \mathrm{H}(R,C_{p^n}) & \longrightarrow & \mathrm{P}(R,C_{p^n}) & \longrightarrow & 1 \\
 & & \downarrow{\scriptstyle j_0} & & \downarrow{\scriptstyle j} & & \downarrow{\scriptstyle \bar{j}} & & \\
1 & \longrightarrow & \mathrm{NB}(S_n,C_{p^n})^{\Gamma_n} & \longrightarrow & \mathrm{Disc}(S_n,p^n)(-1)^{\Gamma_n} & \xrightarrow{\pi''} & \mathrm{Pic}(S_n)[p^n](-1)^{\Gamma_n} & & \\
 & & \downarrow & & \downarrow & & & & \\
 & & \mathrm{Coker}(j_0) & \xrightarrow{\cong} & \mathrm{Coker}(j). & & & &
\end{array}$$

Note that we have suppressed some isomorphisms from the notation, and that π'' is the restriction of the map π' which was defined in the proof of **0**.5.4. From the above diagram one sees by the snake lemma applied to the two middle rows that $\bar{j}$ is injective. Let us for a moment replace the term $\mathrm{Pic}(S_n)[p^n](-1)^{\Gamma_n}$ by $Q =$ image of π''. One does get a map j': $\mathrm{P}(R,C_{p^n}) \to Q$ induced by $\bar{j}$. The snake lemma now gives an exact sequence

$$\mathrm{Coker}(j_0) \xrightarrow{\cong} \mathrm{Coker}(j) \longrightarrow \mathrm{Coker}(j') \longrightarrow 0,$$

in other words: $\mathrm{Coker}(j') = 0$, i.e. $\mathrm{Im}(\bar{j}) = Q = \mathrm{Im}(\pi'')$. After these preparations, we state

Theorem 3.6. *The homomorphism $\bar{j}$ in the above diagram is bijective, i.e. we have* $\mathrm{P}(R,C_{p^n}) \cong \mathrm{Pic}(S_n)[p^n](-1)^{\Gamma_n}$.

Proof. By the preceding remarks, it suffices to show that π'' is surjective. This is, however, a nontrivial fact, since taking Γ_n-invariants is not right exact. There doesn't seem to be a simple cohomological argument giving surjectivity of π''. We give a detailed proof only for the case that S_n has an artinian quotient ring K, so that all invertible S_n-modules are isomorphic to ideals of S_n (e.g., S_n a domain, or reduced noetherian). This amply suffices for later applications. For the general case, see the comment at the end of the proof.

Recall the setting of 2.10: β is a generator of Γ_n, $w \in \mathbb{Z}$ is a lifting of $\omega(\beta)$ such that $1-w^m$ ($m = |\Gamma_n|$) is precisely divisible by p^n. Finally, $\xi = \sum_{i=0}^{m-1} w^i \cdot \beta^{-i}$. We have $(1-w\beta^{-1})\xi = 1-w^m = qp^n$ with $q \in \mathbb{Z}$ prime to p.

For any invertible S_n*-submodule* $P \subset K$ and $\gamma \in \Gamma_n$, one has $P^\gamma \cong \gamma(P) \subset K$ canonically (proof of Prop. 2.4). The Γ_n-action on discriminant modules also simplifies: if $P \subset K$ is invertible, and μ: $P^{\otimes p^n} \cong P^{p^n} \longrightarrow S_n$ an isomorphism, then $(P,\mu)^\gamma \cong (\gamma(P), \gamma\mu\gamma^{-1})$. We extend the operation to $\mathbb{Z}[\Gamma_n]$ as follows: For every invertible $P \subset K$, one defines $P^{-1} = \{x \in K \mid xP \subset S_n\}$, and one then has $P \cdot P^{-1} = S_n$. Hence P^c makes sense for all $c \in \mathbb{Z}$ since the set of invertible submodules of K is a group under multiplication. For $z = \sum c_\gamma \cdot \gamma \in \mathbb{Z}[\Gamma_n]$ and $P \subset K$ an invertible S_n-module, we define

$$P^z = \prod_{\gamma \in \Gamma_n} \gamma(P)^{c_\gamma}.$$

Similarly, if (P,μ) is a discriminant module, i.e. μ: $PP^n \longrightarrow S_n$ is multiplication with a unit u of K, we let $(P,\mu)^z = (P^z, \mu^z)$, where μ^z is multiplication by $u^z = \prod \gamma(u)^{c_\gamma}$. One checks that

$$(P,\mu)^{zz'} = ((P,\mu)^z)^{z'} = ((P,\mu)^{z'})^z \qquad \text{and}$$
$$(P,\mu)^{z+z'} = (P,\mu)^z \cdot (P,\mu)^{z'},$$

for each p^n-discriminant module (P,μ) with $P \subset K$, and all z, z' $\in \mathbb{Z}[\Gamma_n]$. It is important to have equalities here, not just canonical isomorphisms.

Now we can show that π'' is onto. Let $[P] \in \mathrm{Pic}(S_n)[p^n](-1)^{\Gamma_n}$. We may assume that P is a submodule of K. The fact that $[P]$ is stable under Γ_n (the twist taken into account) tells us that $[P]^{1-w^{-1}\beta}$ is trivial, i.e. $P^{1-w^{-1}\beta}$ is a principal fractional ideal. Let h: $P^{1-w^{-1}\beta} \rightarrow S_n$ be an isomorphism. What we want is a preimage of $[P]$ under π''. Let $Q = P^q$; we shall construct a discriminant module (Q,μ) which lies in $\mathrm{Disc}(S_n,p^n)(-1)^{\Gamma_n}$. Then $[Q] = \pi''[(Q,\mu)]$, and we are done because $\mathrm{Disc}(S_n,p^n)$ is p-primary, and q is prime to p.

The construction of μ is tricky; take μ to be the map

$$QP^n = P^qP^n = P^{(1-w^{-1}\beta)\xi} \xrightarrow{h^\xi} S_n.$$

We have to show that (Q,μ) is fixed under the (-1)-twisted action of β, i.e. that $(Q,\mu)^{1-w^{-1}\beta}$ is isomorphic to the trivial discriminant module (S_n, id_{S_n}). Define f: $Q^{1-w^{-1}\beta} \longrightarrow S_n$ as follows:

$$Q^{1-w^{-1}\beta} = P^{(1-w^{-1}\beta)\cdot q} \xrightarrow{h^q} S_n.$$

We now must check commutativity of the triangle

$$\begin{array}{ccc} (Q^{1-w^{-1}\beta})P^n & \xrightarrow{fP^n} & S_n \\ & \mu^{1-w^{-1}\beta} \searrow \quad \swarrow\!\!\!= & \\ & S_n. & \end{array}$$

If we plug in the definitions of f and μ, the map fP^n becomes h^{qP^n}, and the other map becomes $h^{\xi(1-w^{-1}\beta)}$. Since $qp^n = \xi(1-w^{-1}\beta)$, the diagram indeed commutes, hence f is a morphism of discriminant modules, q.e.d.

A few comments on the general case: The first possibility is to redo the whole argument without the benefit of an artinian quotient ring K of S_n. The quantity $(P,\mu)^z$ still can be defined, but the equalities in the formulas above become canonical isomorphisms. The final triangle in the above proof commutes only up to several such isomorphisms, and one has an awkward time making the argument precise. (See Greither (1988), p.24.)

Another proof goes as follows: By a reduction modulo nilpotents, show that the theorem holds for all R (and S_n) which are of finite type over $\mathbb{Z}[p^{-1}]$. Then one can reduce the theorem to this special case by a standard finiteness argument. A lot of details have to be checked.

Let us pause briefly to sum up what we know about cyclotomic descent now:

a) There is a map j_0: $\mathrm{NB}(R, C_{p^n}) \longrightarrow \mathrm{NB}(S_n, C_{p^n})^{\Gamma_n}$ with kernel and cokernel cyclic of the same (finite and explicitly given) order; (3.3)

b) same as a) for H instead of NB; (3.4)

c) for the factor group P = H/NB, there is an isomorphism

$$\bar{j}:\ \mathrm{P}(R, C_{p^n}) \longrightarrow P(S_n, C_{p^n})^{\Gamma_n}. \quad (3.6)$$

Corollary 3.7. *If all groups involved are finite, one has*

a) $|\mathrm{NB}(R, C_{p^n})| = |\mathrm{NB}(S_n, C_{p^n})^{\Gamma_n}| = |(S_n^*/p^n)(-1)^{\Gamma_n}|$;

b) $|\mathrm{H}(R, C_{p^n})| = |\mathrm{H}(S_n, C_{p^n})^{\Gamma_n}|$.

These results suggest the question whether $\mathrm{NB}(R, C_{p^n})$ is actually isomorphic as abelian group to $\mathrm{NB}(S_n, C_{p^n})^{\Gamma_n}$ (and the same question for H, of course). The answer to both questions is yes.

Theorem 3.8. *a) The groups* $\mathrm{NB}(R, C_{p^n})$ *and* $\mathrm{NB}(S_n, C_{p^n})^{\Gamma_n}$ *are isomorphic.*

b) The groups $\mathrm{H}(R, C_{p^n})$ *and* $\mathrm{H}(S_n, C_{p^n})^{\Gamma_n}$ *are isomorphic.*

Proof. We prove only a); the argument for b) is analogous. To begin with, note that $\mathrm{H}(R, C_{p^n})$ is killed by p^n. More precisely, if $A \in \mathrm{H}(R, C_{p^n})$ is connected, then $[A]$ has order exactly p^n by Lemma 1.8; if A is not connected, then it comes from $\mathrm{H}(R, C_{p^s})$ for some $s < n$, and we are done by induction.

The case $n \le n_0(R)$ is trivial: here $|\Gamma_n| = |\Gamma_1|$ is prime to p, hence g.c.d.$(|\Gamma_n|, p^n) = 1$, and j_0 is already an isomorphism. Assume henceforth $n_0 < \infty$ and $n > n_0$. The idea is the following: Find subgroups $D_1 \subset N_1 = \mathrm{NB}(R, C_{p^n})$, $D_2 \subset N_2 = \mathrm{NB}(S_n, C_{p^n})^{\Gamma_n}$ such that:

(*i*) D_1 and D_2 are cyclic of order p^n, and

(*ii*) j_0 induces an isomorphism $N_1/D_1 \longrightarrow N_2/D_2$.

This will suffice, since by elementary group theory D_i is a direct summand of N_i $(i = 1, 2)$.

For D_1 we take the cyclic group generated by $A_n(R)$, the n-th cyclotomic p-extension of R (see 1.8 and preceding discussion). For D_2 we take the cyclic group generated by $A_n(S_n)$, the n-th cyclotomic extension of S_n. Since $n > n_0(R)$, the invariant $n_0(S_n)$ equals n, hence $A_n(S_n)$ is the p-part of the p^{2n}-th cyclotomic extension of S_n, which happens to be S_{2n}. Since $[S_{2n}:S_n] = p^n$, $A_n(S_n)$ is just S_{2n}

itself, whereas $A_n(R)$ is the p-part of S_{n+n_0}. By 2.4, $A_n(R)/R$ and $A_n(S_n)/S_n$ possess normal bases. Also by 2.4, $j_0(A_n(R))$ is a Kummer extension $S_n(p^n;z)$ with z a primitive p^{n_0}-th root of unity in S_n. This implies $p^{n_0}\cdot A_n(R) \in \mathrm{Ker}(j_0)$. Since $A_n(R)$ has order p^n, and $|\mathrm{Ker}(j_0)| = \text{g.c.d.}(p^n, |\Gamma_n|) = p^{n-n_0}$, we get that $\mathrm{Ker}(j_0)$ is generated by $p^{n_0}\cdot A_n(R)$. On the other hand, it is clear from 1.6 that $A_n(S_n) = S_{2n} = S_n(p^n;\zeta)$ for some primitive p^n-th root of unity $\zeta \in S_n$. This shows that $j_0(C_1)$ is the subgroup of order p^{n_0} in D_2.

In the proof of 3.4 we constructed an isomorphism δ: $\mathrm{Coker}(j_0) \longrightarrow \mu_{p^n}/m$, $m = |\Gamma_n|$. Obviously $\mu_{p^n}/m = \mu_{p^n}/(\text{g.c.d.}(m,p^n)) = \mu_{p^n}/p^{n-n_0}$. We showed there that ζ maps to a generator of μ_{p^n}/m under δ. Hence the restriction of δ: $D_2 \to \mu_{p^n}/m$ is onto, and by cardinality reasons its kernel is the subgroup of order p^{n_0} in D_2. Putting everything together, we obtain a diagram with two exact rows:

$$
\begin{array}{ccccccccccc}
0 & \longrightarrow & \mathrm{Ker}(j_0) & \longrightarrow & D_1 & \longrightarrow & D_2 & \longrightarrow & \mathrm{Coker}(j_0) & \longrightarrow & 0 \\
 & & \| & & \cap & & \cap & & \| & & \\
0 & \longrightarrow & \mathrm{Ker}(j_0) & \longrightarrow & N_1 & \xrightarrow{j_0} & N_2 & \longrightarrow & \mathrm{Coker}(j_0) & \longrightarrow & 0 \\
 & & & & \downarrow & & \downarrow & & & & \\
 & & & & D_1/N_1 & \longrightarrow & D_2/N_2 & . & & &
\end{array}
$$

By an easy diagram chase one sees that the map $D_1/N_1 \longrightarrow D_2/N_2$ induced by j_0 is bijective, q.e.d.

CHAPTER II

Corestriction and Hilbert's Theorem 90

We begin this chapter with a discussion of corestriction, first in a general setting, then for Galois extensions. After this, we obtain a theorem of type "Hilbert 90" for the group $\mathrm{NB}(S_n, C_{p^n})$ (notations as in the last chapter). As a consequence we get a description of $\mathrm{NB}(R, C_{p^n})$ as a *factor* group of $\mathrm{NB}(S_n, C_{p^n})$. Note that in I §3 we essentially derived a description of $\mathrm{NB}(R, C_{p^n})$ as a *subgroup* of $\mathrm{NB}(S_n, C_{p^n})$, up to a finite term $\mathrm{Ker}(j_0)$.

It is interesting to note that Merkurjev (1986) proved a similar Hilbert 90 theorem for the functor K_2 on fields (which is certainly much more difficult than the result presented in this chapter).

§1 Corestriction

We first study corestriction of algebras, a concept due to Riehm (1970). Cf. also Wenninger (1991). Let R be any commutative ring.

Let T/R be an arbitrary Γ-Galois extension (Γ a finite group), and ALG_R (ALG_T) be the category of commutative R-algebras (T-algebras, respectively). One defines a functor $\mathrm{Cor} = \mathrm{Cor}_{T/R}\colon \mathrm{ALG}_T \longrightarrow \mathrm{ALG}_R$ as follows: Recall that for $\gamma \in \Gamma$ and $B \in \mathrm{ALG}_T$, B^γ is just B, but with T acting via γ^{-1} on it. For $B \in \mathrm{ALG}_T$, we first let

$$\tilde{B} = \bigotimes_{\gamma\in\Gamma} B^\gamma \qquad \text{(the tensor product taken over } T\text{)},$$

and we define a descent datum $(\Phi_\delta)_{\delta\in\Gamma}$ on $\tilde{B}$ by

$$\Phi_\delta(\ldots\otimes b_\gamma \otimes\ldots) = \ldots\otimes b_{\delta^{-1}\gamma}\otimes\ldots \qquad (\delta\in\Gamma,\ b_\gamma \in \tilde{B} \text{ for all } \gamma\in\Gamma)$$

or more compactly: $\Phi_\delta(\bigotimes_\gamma b_\gamma) = \bigotimes_\gamma c_\gamma$ with $c_\gamma = b_{\delta^{-1}\gamma}$ for all $\gamma \in \Gamma$. One checks without trouble that Φ_δ is δ-linear and $\Phi_\delta\Phi_{\delta'} = \Phi_{\delta\delta'}$ ($\delta, \delta' \in \Gamma$). Finally one lets

$$\mathrm{Cor}(B) = \tilde{B}^{\{\Phi_\delta \mid \delta\in\Gamma\}} \qquad \text{(fixed ring of all } \Phi_\delta\text{)}.$$

By Galois descent (0 §7), $\mathrm{Cor}(B)$ is an R-algebra, and $T\otimes_R \mathrm{Cor}(B)$ is canonically isomorphic to $\tilde{B} = \bigotimes_{\gamma\in\Gamma} B^\gamma$.

It is obvious how to define Cor also for T-algebra morphisms, so we obtain a functor Cor: $\mathrm{ALG}_T \longrightarrow \mathrm{ALG}_R$.

Let j: $\mathrm{ALG}_R \longrightarrow \mathrm{ALG}_T$ be the base change functor $A \longmapsto T\otimes_R A$. As a first property of Cor, we then have

Remark 1.1. *For $B \in \mathrm{ALG}_T$ we have $(j\circ\mathrm{Cor})(B) \cong \bigotimes_{\gamma\in\Gamma} B^{\gamma}$.*

Proof. By construction, $j(\mathrm{Cor}(B)) = T\otimes_R \mathrm{Cor}(B) = \tilde{B} = \bigotimes_{\gamma\in\Gamma} B^{\gamma}$.

Further general properties are listed in the following lemma:

Lemma 1.2. *a) For any R-algebra S, and $T_S = S\otimes_R T$, we have $\mathrm{Cor}_{T_S/S}(B_S) \cong (\mathrm{Cor}_{T/R}(B))_S$, i.e. corestriction commutes with base extension of T/R.*

b) If $\Gamma' \subset \Gamma$ is a normal subgroup, $T' = T^{\Gamma'}$, then we have, up to a canonical isomorphism

$$\mathrm{Cor}_{T/R} \quad = \quad \mathrm{Cor}_{T'/R} \circ \mathrm{Cor}_{T/T'} \, .$$

The *proofs* are straightforward. For b), see also Kersten (1990), Satz 18.4.

If the T-algebra B carries another structure, Cor(B) will often inherit that structure. Specifically, if B is G-Galois, then A = Cor(B) inherits the G-action because the descent datum $(\Phi_\delta)_{\delta\in\Gamma}$ is G-equivariant. But A has no chance of being itself G-Galois over R since it is too big: one easily calculates that $[A:R] = [B:T]^{|\Gamma|}$. Hence the construction has to be modified; this works properly only for G abelian, so we henceforth assume this. Let μ: $\prod_{\gamma\in\Gamma} G \longrightarrow G$ the multiplication map (a homomorphism!). If we let each factor G operate on one tensor factor B^{γ}, the algebra $\tilde{B} = \bigotimes_{\gamma\in\Gamma} B^{\gamma}$ becomes a $\prod_{\gamma\in\Gamma} G$-Galois extension of T, and we may form the G-Galois extension $B^{\#} = \mu^{*}B$ of T. As we know, $B^{\#}$ is just the fixed ring of $\mathrm{K} = \mathrm{Ker}(\mu)$. Now the descent maps $\tilde{\Phi}_\delta$ ($\delta \in \Gamma$) do not commute with the elements of K, but one can check that at least $\Phi_\delta \mathrm{K} = \mathrm{K}\Phi_\delta$ for $\delta \in \Gamma$. Hence the restriction of Φ_δ to $B^{\#}$ is well-defined (and one again gets a descent datum), so we may define

$$\mathrm{cor}(B) \; = (B^{\#})^{\{\Phi_\delta | \delta\in\Gamma\}} \quad (= \; \mathrm{Cor}(B) \cap B^{\#}).$$

Again by descent, cor(B) is a G-Galois extension of R (since $B^{\#}/T$ is G-Galois). Note that $B^{\#}$ is the Harrison product of all B^{γ} (and *not* the tensor product). From this we immediately get (cf. 1.1):

Remark 1.3. *For $B \in \mathrm{H}(T, C_{p^n})$, $(j\circ\mathrm{cor})(B) = \prod_{\gamma\in\Gamma} B^{\gamma} \in \mathrm{H}(T, C_{p^n})$*

There is another way of stating this: If $\mathrm{N}_\Gamma = \sum_{\gamma\in\Gamma}\gamma$ is the norm element of the group ring $\mathbb{Z}[\Gamma]$, then $j\circ\mathrm{cor}$ = multiplication by N_Γ in the $\mathbb{Z}[\Gamma]$-module $\mathrm{H}(T, C_{p^n})$.

In the situation of Chapter I, we have also a corestriction of Galois extension with normal basis: Suppose $p^{-1} \in R$ (p an odd prime), R connected, S_n the p^n-th cyclotomic extension of R, and Γ_n the Galois group of S_n over R.

Lemma 1.5. *With* $T = S_n$, $\Gamma = \Gamma_n$ *in the above notation,* cor: $H(S_n, C_{p^n}) \longrightarrow H(R, C_{p^n})$ *maps* $NB(S_n, C_{p^n})$ *to* $NB(S_n, C_{p^n})$.

Proof. Let $B \in NB(S_n, C_{p^n})$. By I 2.1 it suffices to show that $j\mathrm{cor}(B)$ has normal basis over S_n. Now $j\mathrm{cor}(B) = \prod_{\gamma \in \Gamma_n} B^\gamma$, and each B^γ is in $NB(S_n, C_{p^n})$ since NB is a subgroup functor of H. This also shows that $j\mathrm{cor}(B) \in NB(S_n, C_{p^n})$, q.e.d.

We now return to an arbitrary Γ-extension T/R. Let us, however, assume T (and R) connected. In **0** §3 we gave a description of $H(R,G)$ as $\mathrm{Hom}_{cont}(\Psi_R, G)$ for G finite abelian, where Ψ_R is the automorphism group of R^{sep}. Furthermore, as shown in the proof of I 3.3, j (= base change with T over R) corresponds to the restriction, i.e. the following diagram commutes:

$$\begin{array}{ccc} H(R,G) & \xrightarrow{j} & H(T,G) \\ \downarrow & & \downarrow \\ \mathrm{Hom}_{cont}(\Psi_R, G) & \xrightarrow{res} & \mathrm{Hom}_{cont}(\Psi_T, G) \\ \| & & \| \\ H^1(\Psi_R, G) & \xrightarrow{res} & H^1(\Psi_T, G) . \end{array}$$

Note here that $\Psi_T \subset \Psi_R$: choose a copy of T inside R^{sep} and identify Ψ_T with $\mathrm{Aut}(R^{sep}/T)$.

The question arises: what is the description of $\mathrm{cor}_{T/R}$, viewed as a map $H^1(\Psi_T, G) \longrightarrow H^1(\Psi_R, G)$? Here, as in the diagram, G is a trivial module over Ψ_T and Ψ_R. Of course there is already a well-known map called corestriction in group cohomology. We recall its definition (cf. Merkurjev (1986)): Let Ψ' be a normal subgroup of finite index of the profinite group Ψ, and let M be a discrete trivial Ψ'-module. Then *cor*: $H^1(\Psi', M) \longrightarrow H^1(\Psi, M)$ is defined as follows: Let $\Gamma = \Psi/\Psi'$, and $\{\psi_\gamma \mid \gamma \in \Gamma\}$ be a system of representatives of Ψ mod Ψ'. For f: $\Psi' \longrightarrow M$ continuous, define

$$cor(f)(\sigma) = \sum_{\gamma \in \Gamma} f(\psi_\gamma \sigma (\psi_{\gamma\bar{\sigma}})^{-1}).$$

One also knows that $res \circ cor$ = multiplication by N_Γ on $H^1(\Psi', M)$ (the factor group Γ operates on $H^1(\Psi', M)$, cf. proof of I 3.3). So one is led to conjecture:

Proposition 1.6. *The diagram*

$$\begin{array}{ccc} H(T,G) & \xrightarrow{\mathrm{cor}_{T/R}} & H(R,G) \\ \downarrow & & \downarrow \\ H^1(\Psi_T, G) & \xrightarrow{cor} & H^1(\Psi_R, G) \end{array}$$

commutes. (We again identify Ψ_T *with a subgroup of* Ψ_R, *such that* $\Gamma = \Psi_R/\Psi_T$.*)*

Proof. This is complicated, and we shall first treat the case where all rings needed in the course of the proof are connected. Let us at the beginning explain the connection between Galois descent and H^1.

Suppose $C \in H(T,G)$ is connected and carries a descent datum $(\Phi_\gamma)_{\gamma\in\Gamma}$. Assume without loss that $T, C \subset R^{sep}$. Let $A =$ the fixed ring of all Φ_γ. Then $C = T \otimes_R A$ is $\Gamma\times G$-Galois over R, and Φ defines a homomorphism $\Gamma \longrightarrow \mathrm{Aut}(C/R) = \Psi_R/\Psi_C$ (with $\Psi_C = \mathrm{Aut}(R^{sep}/C)$). The isomorphism $\beta = \nu^{-1}$: $H(R,G) \longrightarrow H^1(\Psi_R, G)$ (see proof of **0**.3.6) has a simple description for any connected $A' \in H(R,G)$: $\beta(A')(\sigma) =$ the element g of G that acts like σ on A' ($\sigma \in \Psi_R$). In particular we have

$$(*) \qquad \beta(A)(\sigma) = \beta(B)(\Phi_{\bar\sigma}^{-1}\cdot\sigma), \text{ with } \bar{\ }: \Psi_R \to \Gamma \text{ the canonical map.}$$

The point is here: $\beta(B)$ is only defined on Ψ_T, but $\Phi_{\bar\sigma}^{-1}\cdot\sigma$ is identity on T for all $\sigma \in \Psi_R$. To save space, we abused notation: For $\Phi_{\bar\sigma}^{-1}$, one has to read: any lift of $\Phi_{\bar\sigma}^{-1}$ to Ψ_R.

Suppose now $B \in H(T,G)$ such that $\tilde B = \bigotimes_{\gamma\in\Gamma}$ is still connected. Then $\tilde B$ embeds into R^{sep}: map B^γ to $\psi_\gamma(B)$, where ψ_γ is some lift of γ to Ψ_R; the resulting map f: $\tilde B \to R^{sep}$ must be injective by **0**.3.7. The definition of the descent datum Φ_δ in the construction of $\mathrm{Cor}(B)$ shows that (if we consider Φ_δ as an automorphism of $\mathrm{Im}(f)$) Φ_δ maps $\psi_\gamma(B)$ to $\psi_{\delta\gamma}(B)$ via $\psi_{\delta\gamma}\cdot\psi_\gamma^{-1}$ (*not* via ψ_δ, this would not even result in a descent datum). We now calculate for $h \in \Psi$:

$$\beta(\tilde B)(h) = \left(\beta(B^\gamma)(h)\right)_{\gamma\in\Gamma} \in G^\Gamma \qquad \text{(use the above description of } \beta\text{),}$$

hence by functoriality of β and since $B^* = \mu^*\tilde B$:

$$\beta(B^*)(h) = \Sigma_{\gamma\in\Gamma}\beta(B^\gamma)(h) \in G.$$

Moreover, again by the above description of β,

$$\beta(B^\gamma) = \beta(B)(\psi_\gamma^{-1}h\psi_\gamma).$$

Let now $\sigma \in \Psi_R$. We just calculated that $\Phi_{\bar\sigma}$ maps B^γ to $B^{\bar\sigma\gamma}$ as $\psi_{\sigma\gamma}\cdot\psi_\gamma^{-1}$, hence $\sigma^{-1}\Phi_{\bar\sigma}$ acts on B^γ as $\sigma^{-1}\psi_{\bar\sigma\gamma}\cdot\psi_\gamma^{-1}$. Let $A = \mathrm{cor}(B)$. Then:

$$\begin{aligned}
\beta(A)(\sigma) &= \beta(B^*)(\Phi_{\bar\sigma}^{-1}\,\sigma)\\
&= \Sigma_{\gamma\in\Gamma}\,\beta(B^\gamma)(\Phi_{\bar\sigma}^{\ -1}\sigma)\\
&= \Sigma_{\gamma\in\Gamma}\,\beta(B^\gamma)(\psi_\gamma\cdot\psi_{\bar\sigma\gamma}^{-1}\cdot\sigma)\\
&= \Sigma_{\gamma\in\Gamma}\beta(B)(\psi_\gamma^{-1}\,\psi_\gamma\cdot\psi_{\bar\sigma\gamma}^{-1}\cdot\sigma\cdot\psi_\gamma)\\
&= \Sigma_{\gamma\in\Gamma}\beta(B)(\psi_{\bar\sigma\gamma}^{-1}\cdot\sigma\cdot\psi_\gamma)
\end{aligned}$$

Now we pass to the "converse" system of representatives $\psi'_\delta = (\psi_{\delta^{-1}})^{-1}$ of Ψ_R mod Ψ_T, and by abuse of notation, we omit the ' again. The last expression then becomes

$$
\begin{aligned}
&= \Sigma_{\gamma\in\Gamma}\beta(B)(\psi_{(\bar{\sigma}\gamma)^{-1}}\cdot\sigma\cdot(\psi_{\gamma^{-1}})^{-1}) \\
&= -\Sigma_{\gamma\in\Gamma}\beta(B)(\psi_{\gamma^{-1}}\cdot\sigma^{-1}\cdot(\psi_{\gamma^{-1}\bar{\sigma}^{-1}})^{-1}) \\
&= -\,cor(\beta(B))(\sigma^{-1}) \\
&= cor(\beta(B))(\sigma), \qquad \text{q.e.d.}
\end{aligned}
$$

In the general case (B not necessarily connected), it seems that there is no easy reduction argument to the connected case, since $\tilde{B}$ may well be unconnected even though B is connected. It seems preferable to argue as follows: If a G-Galois extension S of T comes by functoriality from a connected G'-extension S', with $G' \subset G$ a subgroup, then the algebra S is a product of copies of S_0, and the description of $\beta(S)$: $\Psi_T \to G$ is as follows: $\beta(S)(\sigma)$ is the element of G (or of G_0) which acts like σ on each factor S_0 in S. (Here one has to think of S_0 as a subalgebra of R^{sep}.) With this rephrasing of the description of β, the previous argument still works.

Remark. This proposition is "nice to know", but in the proofs in §3 it is only used in minor places. It would be nice to use 1.5 systematically, and prove results about corestriction of Galois extensions via Galois cohomology, but it seems that this is not so easy. In fact, we only understand well corestriction of extensions *with normal basis* as the reader will see in §3; and the group cohomological description is not well suited to deal with normal bases. Let us point out one nice consequence of 1.5:

Corollary 1.6. *For all* $\beta \in \Gamma$, $j\circ$(*multiplication by* $1-\beta$) *is the zero map.*

Proof. This holds by 1.5 since the analogous statement in group cohomology is true (cf. Merkurjev (1986)). One could also give a direct argument, avoiding the use of 1.5.

§2 Lemmas on group cohomology

We only use rudiments of group cohomology (nothing beyond H^1). Let p be an odd prime, $n \geq 1$, and Γ a group with a fixed embedding $\omega: \Gamma \to (\mathbb{Z}/p^n)^*$. For every Γ-module M with $p^n M = 0$, we defined in I §2 twisted modules $M(k)$ ($k \in \mathbb{Z}$). (The case $k = -1$ was used extensively in the sequel.)

Let us quickly recall the definition of the zeroth and first Tate cohomology group. We pick a generator β of Γ. Then

$$\begin{aligned} \hat{H}^0(\Gamma, M) &= M^\Gamma / N_\Gamma M \quad (N_\Gamma = \textstyle\sum_{\gamma\in\Gamma}\gamma) \text{ "fixed elements mod. norms"} \\ &= M^{1-\beta}/N_\Gamma M \quad \text{(the exponent means: kernel of . . .);} \\ H^1(\Gamma, M) &= \hat{H}^1(\Gamma, M) = M^{N_\Gamma}/(1-\beta)M \\ &\qquad \text{"kernel of the norm modulo image of } 1-\beta\text{".} \end{aligned}$$

Moreover we recall that $\omega(\beta)$ has a lifting $w \in \mathbb{Z}$ such that $1 - w^m$ is divisible exactly by the power p^n ($m = |\Gamma|$), and $\xi = \sum_{i=0}^{m-1} w^i \beta^{-i}$. One has (I 2.9, I 2.10)

$$(1 - w\beta^{-1})\cdot\xi = 1 - w^m = p^n q, \qquad q \text{ prime to } p.$$

Lemma 2.1. *If N is a Γ-module without p-torsion, and $M = N/p^n N$, then both $H^0_T(\Gamma, M(-1))$ and $H^1(\Gamma, M(-1))$ are zero.*

Proof. Note first: N_Γ acts on $M(-1)$ just as ξ acts on M; $1-\beta$ acts on $M(-1)$ just as $1 - w^{-1}\beta$ acts on M.

$$\begin{aligned} \text{Hence: } \hat{H}^0(\Gamma, M(-1)) &= M^{(1-w^{-1}\beta)}/\xi M; \\ H^1(\Gamma, M(-1)) &= M^\xi/(1-w^{-1}\beta). \end{aligned}$$

The argument for the vanishing of H^0_T is almost the same as for H^1:

For $\hat{H}^0$: Suppose $x \in N$, $\bar{x} \in M^{(1-w^{-1}\beta)}$. Then $(1-w^{-1}\beta)x = p^n y$ for some $y \in N$. Hence $q\cdot(1-w^{-1}\beta)x = qp^n y = (1-w^{-1}\beta)\xi y$, so $(1-w^{-1}\beta)(qx - \xi y) = 0$. Multiplying this with ξ, we get $p^n q\cdot(qx - \xi y) = 0$, whence by hypothesis $q\cdot(qx - \xi y) = 0$, i.e. $q^2 x \in \xi N$, and $q^2\bar{x} \in \xi M$. Since $p^n M = 0$ and q is prime to p, also $\bar{x} \in \xi M$.

For H^1: Suppose $x \in N$, $\bar{x} \in M^\xi$. Then $\xi x = p^n y$ for some $y \in N$. Hence $q\cdot\xi x = qp^n y = (1-w^{-1}\beta)\xi y$, so $\xi\cdot(qx - (1-w^{-1}\beta)y) = 0$. Multiplying this with $(1-w^{-1}\beta)$, we get $p^n q\cdot(qx - (1-w^{-1}\beta)y) = 0$, whence by hypothesis $q\cdot(qx - (1-w^{-1}\beta)y) = 0$, $q^2 x \in (1-w^{-1}\beta)N$, and $q^2\bar{x} \in (1-w^{-1}\beta)M$. Since $p^n M = 0$ and q is prime to p, also $\bar{x} \in (1-w^{-1}\beta)M$.

Remark. One may get another proof by using Thm. IX 9 and its corollary from Serre (1968) (note that $M = \mathbb{Z}/p^n(-1)\otimes_{\mathbb{Z}} N$ and $\mathbb{Z}/p^n(-1) \cong \mathrm{Hom}(\mu_{p^n}, \mathbb{Q}_p/\mathbb{Z}_p)$). Furthermore, our proof also works for twists by any $k \in \mathbb{Z} - p\mathbb{Z}$ instead of -1.

Corollary 2.2. *If N is a Γ-module without p-torsion, and $0 \to A \to B \to N \to 0$ is a short exact sequence of Γ-modules, then the two sequences*

$$0 \longrightarrow (A/p^n)(-1)^{\Gamma} \longrightarrow (B/p^n)(-1)^{\Gamma} \longrightarrow (N/p^n)(-1)^{\Gamma} \longrightarrow 0$$

and

$$0 \longrightarrow (A/p^n)(-1)_{\Gamma} \longrightarrow (B/p^n)(-1)_{\Gamma} \longrightarrow (N/p^n)(-1)_{\Gamma} \longrightarrow 0$$

are again exact, where $(.)_{\Gamma}$: $X \longmapsto X_{\Gamma} = X/(1-\beta)X$ is the functor of taking the Γ-coinvariants of a Γ-module. (Again, any $k \in \mathbb{Z}-p\mathbb{Z}$ may replace -1.)

Proof. The sequence $0 \to A/p^n \to B/p^n \to N/p^n \to 0$ is exact since N has no p-torsion (in higher language, $\mathrm{Tor}_{\mathbb{Z}}(N, \mathbb{Z}/p^n)$ vanishes). Twisting with -1 does not affect the exactness. Thus it remains to see in the first sequence that π: $(B/p^n)(-1)^{\Gamma} \longrightarrow (N/p^n)(-1)^{\Gamma}$ is onto. But $\hat{\mathrm{H}}^0(\Gamma, (N/p^n)(-1)) = 0$ by 2.1, hence $(N/p^n)(-1)^{\Gamma} = \xi\cdot(N/p^n)$. It is now obvious that $\xi\cdot(B/p^n) \longrightarrow \xi\cdot(N/p^n)$ is onto, and it is also clear that $\xi\cdot(B/p^n) \subset (B/p^n)(-1)^{\Gamma}$ (which is the annihilator of $(1-w^{-1}\beta)$ in B/p^n). Hence π is onto.

For the second sequence, we need injectivity of $(A/p^n)(-1)_{\Gamma} \longrightarrow (B/p^n)(-1)_{\Gamma}$. Since taking [co-]invariants of Γ-modules is just taking the [co-]kernel of $1-\beta$, we see by the snake lemma that the claimed injectivity is equivalent to the surjectivity of $(B/p^n)(-1)^{\Gamma} \longrightarrow (N/p^n)(-1)^{\Gamma}$ which we have just proved, q.e.d.

The most important application takes place in the following situation: R is connected and contains p^{-1}, S_n is the p^n-th cyclotomic extension of R, $\Gamma = \Gamma_n$, and $n \geq n_0(R)$. Then $\mu_{p^n} = \mu_{p^n}(S_n)$ is the group of p-power roots of unity in S_n, i.e. the p-power torsion in S_n^*. This implies that $N = S_n^*/\mu_{p^n}$ has no p-torsion. Of course, $M = N/p^n = (S_n^*/p^n)/\mathrm{Im}(\mu_{p^n})$, and $\mathrm{Im}(\mu_{p^n})$ is an isomorphic copy of μ_{p^n}. We now obtain from Lemma 2.1:

Corollary 2.3. *Under the above hypotheses, the Γ_n-module $M' = (S_n^*/p^n)/\mathrm{Im}(\mu_{p^n})(-1)$ satisfies $\hat{\mathrm{H}}^0(\Gamma_n, M') = \mathrm{H}^1(\Gamma_n, M') = 0$.*

As proved in I§2, $(S_n^*/p^n)(-1)$ is Γ_n-isomorphic to $\mathrm{NB}(S_n, C_{p^n})$ via the map i from Kummer theory (**0**§5). Moreover, if ζ_n denotes an appropriate generator of μ_{p^n}, then $i(\zeta_n)$ is just the C_{p^n}-extension S_{2n}/S_n by the proof of I 2.4 (with ground ring S_n). In other words, $i(\zeta_n)$ is the n-th cyclotomic p-extension $A_n(S_n)$ of S_n. Hence:

Corollary 2.4. *Under the same hypotheses as in 2.3, the Γ_n-module $M'' = \mathrm{NB}(S_n, C_{p^n})/\langle A_n(S_n)\rangle$ has trivial zeroth and first Tate cohomology.*

As we shall see, this result is the key to our theorems about corestriction in the next section. As usual, the cyclotomic extensions need a separate treatment.

§3 "Hilbert 90": The kernel and the image of the corestriction

We keep the hypotheses: $p^{-1} \in R$ (p an odd prime), S_n and Γ_n as in Chap. I. Recall $n_0 = n_0(R) = \max\{\nu \mid S_\nu = S_1\}$. We shall always assume $n \geq n_0$. Let us further recall that $A_n(R)$, the n-th cyclotomic p-extension of R, is the p-part of the extension S_{n+n_0}/R, and that $A_n(S_n)$ is S_{2n}/S_n. (Here it is superfluous to take the p-part since $[S_{2n}:S_n] = p^n$.) By I 2.4, $A_n(R)/R$ and $A_n(S_n)/S_n$ have normal bases. Note in this context also that we determined the C_{p^n}-action on these extensions only modulo choice of a generator of C_{p^n} (there was no canonical such choice), which suffices for our purpose.

Proposition 3.1. *The map* cor: $\mathrm{NB}(S_n, C_{p^n}) \longrightarrow \mathrm{NB}(R, C_{p^n})$ *induces an isomorphism* $\langle A_n(S_n)\rangle \longrightarrow \langle A_n(R)\rangle$. *Equivalently,* $\mathrm{cor}(A_n(S_n)) = A_n(R)$ *if the* C_{p^n}*-actions are fixed correctly.*

Proof. Note first of all that by I 1.8 the two groups $A_n(S_n)$ and $A_n(R)$ are cyclic of order p^n. We shall use in this proof the description of cor by corestriction in group cohomology (1.5). The C_{p^n}-extensions $A_n(R)/R$ and $A_n(S_n)/S_n$ correspond to continuous homomorphisms $h\colon \Psi_R \to C_{p^n}$ and $f\colon \Psi_{S_n} \to C_{p^n}$, respectively. Both f and h are trivial on the fixed group Σ of S_{2n} since both $A_n(R)$ and $A_n(S_n)$ are contained in S_{2n}. Let $\Psi = \Psi_R/\Sigma$, $\Psi_0 = \Psi_{S_n}/\Sigma$. We have to show: $cor(f') = u\cdot h'$ with $u \in \mathbb{Z}$ prime to p, where $h'\colon \Psi \to C_{p^n}$ and $f'\colon \Psi_0 \to C_{p^n}$ are induced by h and f.

Now since $A_n(S_n)$ is connected, f is surjective; one has $|\Psi_0| = [S_{2n}:R]/[S_n:R] = p^n$, hence the induced map f' is an isomorphism. Moreover Ψ is cyclic. By the next lemma we obtain

$$cor(f')(\sigma) \;=\; f'(\sigma^{[\Psi:\Psi_0]}) \qquad \text{for all } \sigma \in \Psi.$$

A simple cardinality argument shows that $\Psi_0 = \{\sigma^{[\Psi:\Psi_0]} \mid \sigma \in \Psi\}$. Hence $cor(f')$ is onto, and so is h' since also $A_n(R)$ is connected. By cardinality reasons again, both $\mathrm{cor}(f')$ and h' are isomorphisms from Ψ_0 to C_{p^n}, which plainly suffices.

Lemma 3.2. *Let* $\Psi_0 \subset \Psi$ *be finite cyclic groups,* $t = [\Psi:\Psi_0]$, *and* M *any trivial* Ψ*-module. Then for each* $f \in \mathrm{H}^1(\Psi_0, M)$ *and* $\sigma \in \Psi$, *we have* $cor(f)(\sigma) = f(\sigma^t)$.

Proof. Let $s = |\Psi_0|$, τ a generator of Ψ. Then τ^t is a generator of Ψ_0, and the set $\{1, \tau, \ldots, \tau^{t-1}\}$ is a transversal of Ψ modulo Ψ_0. Using the formula given in §1 of this chapter, we get $cor(f)(\tau) = \sum_{i=0,\ldots,t-1} f(\tau^i \cdot \tau \cdot \tau^{-[i+1]})$, where $[j]$ denotes the smallest nonnegative residue of j modulo t, and the latter sum contains only one nonzero term (for $i = t-1$), which equals $f(\tau^t)$, q.e.d.

Now we can state and prove the central result of this chapter, a part of which is an analog of Hilbert's Theorem 90:

Theorem 3.3. *The corestriction* cor: $\mathrm{NB}(S_n, C_{p^n}) \longrightarrow \mathrm{NB}(R, C_{p^n})$ *is surjective, and its kernel equals* $(1-\beta)\cdot\mathrm{NB}(S_n, C_{p^n})$ *where* β *is any generator of* Γ_n.

Proof. Let j: $\mathrm{NB}(R, C_{p^n}) \longrightarrow \mathrm{NB}(S_n, C_{p^n})^{\Gamma_n}$ be the base change map (cf. I 3.1.), and recall:

$$j\cdot\mathrm{cor} = \text{multiplication by } N_{\Gamma_n}. \qquad \text{(Lemma 1.3)}$$

Let $D_1 = \langle A_n(R)\rangle \subset \mathrm{NB}(R, C_{p^n})$ and $D_2 = \langle A_n(S_n)\rangle \subset \mathrm{NB}(S_n, C_{p^n})$ as in I 3.9. From there we have: $\mathrm{Ker}(j) \cong \mathrm{Ker}(j|_{D_1}: D_1 \to D_2)$, $\mathrm{Coker}(j) \cong \mathrm{Coker}(j|_{D_1}: D_1 \to D_2)$. These two conditions tell us in particular that $j^{-1}(D_2) = D_1$.

"cor is onto": Take $z \in \mathrm{NB}(R, C_{p^n})$. Then $j(z) \in \mathrm{NB}(S_n, C_{p^n})^{\Gamma_n}$. Since by 2.4, $\mathrm{H}^0_T(\Gamma_n, \mathrm{NB}(S_n, C_{p^n})/C_2) = 0$, we find some $w \in \mathrm{NB}(S_n, C_{p^n})$ with $j(z) \equiv N_{\Gamma_n}\cdot w$ (mod D_2). Using $j\mathrm{cor} = N_{\Gamma_n}$, we see that $z - \mathrm{cor}(w)$ is in $j^{-1}(D_2)$. Hence $z - \mathrm{cor}(w) \in D_1$, and $D_1 \subset \mathrm{Im(cor)}$ by Prop. 3.1, so that we also get $z \in \mathrm{Im(cor)}$.

"$\mathrm{Ker(cor)} = (1-\beta)\cdot\mathrm{NB}(S_n, C_{p^n})$": The inclusion $\supset$ follows from Cor. 1.6. To show the other inclusion, assume $\mathrm{cor}(w) = 0$, $w \in \mathrm{NB}(S_n, C_{p^n})$. Then $j\cdot\mathrm{cor}(w) = N_{\Gamma_n}\cdot w = 0$. By 2.4, $\mathrm{H}^1(\Gamma_n, \mathrm{NB}(S_n, C_{p^n})/D_2) = 0$, so we find $y \in \mathrm{NB}(S_n, C_{p^n})$ with $w \equiv (1-\beta)y$ (mod D_2). When we apply cor to $d = w - (1-\beta)y \in D_2$, we obtain 0 since $\mathrm{cor}(w) = 0$ and also $\mathrm{cor}((1-\beta)y) = 0$. But cor is injective on D_2 by Prop. 3.1, hence $d = 0$, and $w = (1-\beta)y$, q.e.d.

Corollary 3.4. *If R is semilocal (in particular, if R is a field), then the corestriction* cor: $\mathrm{H}(S_n, C_{p^n}) \longrightarrow \mathrm{H}(R, C_{p^n})$ *is onto, with kernel* $(1-\beta)\cdot\mathrm{H}(S_n, C_{p^n})$.

Proof. The rings $R[C_{p^n}]$ and $S_n[C_{p^n}]$ are likewise semilocal, hence their Picard group is zero. Therefore $\mathrm{H}(R, C_{p^n})$ coincides with $\mathrm{NB}(R, C_{p^n})$ (and similarly for S_n), and the corollary immediately follows from Theorem 3.3.

§4 Lifting theorems

In this section, we present lifting theorems for Galois extensions with normal basis. Slightly more special such theorems have been proved by Saltman (1982) [Thm. 2.1, Cor. 5.3] and Kersten - Michaliček (1988) [Cor. 3.11]. The notation is as in §3; let R' be another connected ring with $p^{-1} \in R$.

Theorem 4.1. *Let f: $R' \to R$ be a surjective ring homomorphism whose kernel is contained in* $\mathrm{Ra}(R)$, *the Jacobson radical of* R. *Then the base change map* $\mathrm{H}(f) = R \otimes_{R'} -$: $\mathrm{NB}(R', C_{p^n}) \longrightarrow \mathrm{NB}(R, C_{p^n})$ *is also surjective.*

Proof. Let S_n' be the p^n-th cyclotomic extension of R'. Let us assume first that $R \otimes_{R'} S_n'$ is still connected. Then by I §1, $R \otimes_{R'} S_n'$ is canonically isomorphic to S_n, and Γ_n is the same group as Γ_n'. Denote the resulting epimorphism $S_n' \longrightarrow S_n$ by f_n. By Lemma 1.2, there is a commutative diagram

$$\begin{array}{ccc} \mathrm{NB}(S_n', C_{p^n}) & \xrightarrow{\mathrm{cor}} & \mathrm{NB}(R', C_{p^n}) \\ \mathrm{H}(f_n) \big\downarrow & & \big\downarrow \mathrm{H}(f) \\ \mathrm{NB}(S_n, C_{p^n}) & \xrightarrow{\mathrm{cor}} & \mathrm{NB}(R, C_{p^n}) \end{array},$$

and both horizontal maps are onto by Thm. 3.3. It therefore suffices to show that the base extension map $\mathrm{H}(f_n)$ is onto. But, by Kummer theory, we may replace this map by the canonical map (induced by f_n):

$$(S_n')^*/p^n \longrightarrow S_n^*/p^n,$$

and show this is surjective. Now $\mathrm{Ker}(f_n) = S_n' \cdot \mathrm{Ker}(f) \subset S_n'$ by flatness, $\mathrm{Ker}(f)$ is in the radical of R', and S_n'/R' is finite. Therefore $\mathrm{Ker}(f_n) \subset S_n' \cdot \mathrm{Ra}(R') \subset \mathrm{Ra}(S_n')$ (the second inclusion can easily be seen by recalling that the Jacobson radical is the intersection of all maximal ideals). Hence *any* preimage of a unit under f_n is itself a unit, and f_n: $(S_n')^* \longrightarrow S_n^*$ is surjective, which suffices.

It remains to treat the case that $R \otimes_{R'} S_n'$ is not connected. Let Γ_n' be the automorphism group of S_n' over R'. Then $R \otimes_{R'} S_n' = \iota^* S_n$ where ι: $\Gamma_n \subset \Gamma_n'$ is an injection, and the fixed ring of Γ_n in $R \otimes_{R'} S_n'$ is just $E = R \times \ldots \times R$, the trivial Γ_n'/Γ_n-extension of R. The corestriction may be factored as follows:

$$\mathrm{NB}(R \otimes_{R'} S_n', C_{p^n}) \xrightarrow{\mathrm{cor}_1} \mathrm{NB}(E, C_{p^n}) \xrightarrow{\mathrm{cor}_2} \mathrm{NB}(R, C_{p^n}),$$

where cor_1 is corestriction for the Γ_n-Galois extension $R \otimes_{R'} S_n'$ of E, and cor_2 is corestriction for the Γ_n'/Γ_n-extension E/R.

It is not hard to show directly that cor_2 is onto (actually, $\mathrm{NB}(E, C_{p^n})$ is a direct sum of copies of $\mathrm{NB}(R, C_{p^n})$, and cor_2 is the sum map to $\mathrm{NB}(R, C_{p^n})$). Furthermore, the Γ_n-extension $R \otimes_{R'} S_n$ of $R \times \ldots \times R = E$ is just a cartesian product of $[\Gamma_n':\Gamma_n]$ copies of the extension S_n/R, hence cor_1 is surjective by Thm. 3.3. Thus, the same argument works in the general case.

This theorem has the following variant:

Theorem 4.2. *Let K be a number field, $\mathfrak{p}$ a prime ideal of K. Then the inclusion $K \to K_\mathfrak{p}$ induces a surjection $\mathrm{H}(K, C_{p^n}) \longrightarrow \mathrm{H}(K_\mathfrak{p}, C_{p^n})$.*

Proof. One uses a similar diagram as in the last proof. The p^n-th cyclotomic extension of K ist the field $K(\zeta_n)$ (recall ζ_n this denotes a primitive p^n-th root of 1). It will then be enough to see that $K \to K_\mathfrak{p}$ induces an *epimorphism* α: $K(\zeta_n)^*/p^n \longrightarrow (K_\mathfrak{p} \otimes_K K(\zeta_n))^*/p^n$. Now $K_\mathfrak{p} \otimes_K K(\zeta_n) = \prod_{\mathfrak{P}|\mathfrak{p}} K(\zeta_n)_\mathfrak{P}$. By approximation, to every $y \in (K_\mathfrak{p} \otimes_K K(\zeta_n))^*$ there is an $x \in K(\zeta_n)^*$ such that $u = y/x \equiv 1 \left(\mathrm{mod}\ p^{n+1} \prod_{\mathfrak{P}|\mathfrak{p}} \mathfrak{P}\cdot\mathcal{O}(K(\zeta_n))_\mathfrak{P}\right)$. It is equally well known that all elements of a $\mathfrak{P}$-adic field which are congruent to 1 (mod $\mathfrak{P}\cdot p^{n+1}$) are p^n-th powers (since $p \neq 2$), whence α is indeed surjective.

Remark. The last argument is easily adapted to the case $p = 2$ (take $n+2$ instead of $n+1$), but the surjectivity of cor (which is vital for the proof) is not granted for $p = 2$. Actually, this surjectivity must fail e.g. for $K = \mathbb{Q}$, because 4.2 is just not true for $p = 2$ and $\mathfrak{p}$ dividing 2, as shown by Wang (1948).

We conclude this chapter by a brief discussion of *generic extensions*. The following definition is essentially due to Saltman (1982).

Definition. A *generic C_{p^n}-pair* over the ring R is a pair of R-algebras $(A^\#, R^\#)$ such that:

a) $R^\#$ is of the form $\mathrm{Sym}_R(V)[N^{-1}]$ for some finitely generated projective R-module V and some $N \in \mathrm{Sym}_R(V)$;

b) $A^\# \in \mathrm{NB}(R^\#, C_{p^n})$; and

c) for every $A \in \mathrm{NB}(R, C_{p^n})$, there exists an R-algebra homomorphism $\varphi = \varphi_A$: $R^\# \to R$ such that $A \cong R \otimes_\varphi A^\#$.

Note that $\mathrm{Sym}_R(V)$ is a polynomial ring over R if V is free. (It is quite intentional that we restrict attention to extensions with normal basis in this definition.)

One then has the following existence result for $p \neq 2$, $p^{-1} \in R$, R connected:

Theorem 4.3. *Under the above hypotheses, there exists a generic C_{p^n}-pair over R, and one may choose $V = S_n^\vee$. ($^\vee$ means: R-dual module.)*

Remark. For R a field, this is due to Saltman (loc. cit.). One can actually show that S_n is always free over R, but this would lead us too far afield here.

Proof of **4.3**: Let $V = S_n^\vee$. Consider the norm map $N = \mathrm{N}_{\Gamma_n}\colon S_n \to R$. Since S_n is faithfully flat over R, we have $S_n^* \cap R = R^*$. From this one sees: $u \in S_n$ is a unit iff $N(u)$ is a unit in R.

Let us suppose for a moment that S_n is R-free, and choose a basis. Then N is given by a homogeneous polynomial of degree $m = [S_n:R]$ in m variables. If S_n is not supposed free, then N is an element of the m-th homogeneous piece $\mathrm{Sym}_R(V)_m$ with $V = S_n^\vee$. Let $R^\# = \mathrm{Sym}_R(V)[N^{-1}]$. For any R-algebra T one has:

$$\begin{aligned}\mathrm{Alg}_R(R^\#,T) &= \{\varphi \mid \varphi\colon \mathrm{Sym}_R(V) \to T \ R\text{-algebra hom., } \varphi(N) \text{ invertible}\}\\ &= \{\varphi \mid \varphi\colon S_n^\vee \to T \ R\text{-linear, } \mathrm{Sym}(\varphi)_m(N) \text{ invertible}\}\\ &= \{u \mid u \in S_n \otimes_R T,\ N(u) \text{ invertible}\}\\ &= (S_n \otimes_R T)^*.\end{aligned}$$

(In fancier terms: $\mathrm{Spec}(R^\#)$ is the *Weil restriction* of $\mathbf{G}_m$ from S_n to R.)

One shows that $S_n \otimes_R V$ is S_n-linearly, Γ_n-equivariantly isomorphic to $\bigoplus_{\gamma\in\Gamma_n} S_n X_\gamma$ (the X_γ are the coordinate functions on $S_n \otimes_R S_n \cong S_n^{(\Gamma_n)}$), and the norm N goes to the product $\prod_\gamma X_\gamma$ of the coordinate functions. Hence:

$$\begin{aligned}S_n \otimes_R R^\# &\cong \mathrm{Sym}_{S_n}(S_n \otimes_R V)[N^{-1}]\\ &\cong \mathrm{Sym}_{S_n}(\textstyle\bigoplus_{\gamma\in\Gamma_n} S_n X_\gamma)\big[(\prod X_\gamma)^{-1}\big]\\ &= S_n[X_\gamma]_{\gamma\in\Gamma_n}\big[(\prod X_\gamma)^{-1}\big].\end{aligned}$$

The canonical descent datum on $S_n \otimes_R R^\#$ permutes the X_γ by index shift. Now define:

$$\begin{aligned}B^\# &= (S_n \otimes_R R^\#)(p^n;\, X_1); \qquad \text{"adjoin } p^n\text{-th root of } X_1\text{"}\\ A^\# &= \mathrm{cor}_{S_n\otimes_R R^\#/R^\#}(A^\#).\end{aligned}$$

We claim that $(A^\#, R^\#)$ is a generic pair. Properties a) and b) are clear from the construction. For c), take $A \in \mathrm{NB}(R, C_{p^n})$ and pick by 3.3 some $B \in \mathrm{NB}(S_n, C_{p^n})$ with $\mathrm{cor}_{S_n/R}(B) = A$. Then write $B = S_n(p^n;u)$ $(u \in S_n^*)$ via Kummer theory. Define $\varphi'\colon S_n \otimes_R R^\# \longrightarrow S_n$ by $X_\gamma \longmapsto \gamma(u)$, $\gamma \in \Gamma_n$. Then φ' descends to $\varphi\colon R^\# \to R$. Moreover, $S_n \otimes_{\varphi'} B^\# \cong B$ since $\varphi'(X_1) = u$. By 1.3 we finally get:

$$\begin{aligned}A = \mathrm{cor}_{S_n/R}(B) &\cong \mathrm{cor}_{S_n/R}(S_n \otimes_{\varphi'} B^\#)\\ &\cong R \otimes_\varphi \mathrm{cor}_{S_n\otimes_R R^\#/R^\#}(B^\#) \quad \text{by Lemma 1.2 a)}\\ &= R \otimes_\varphi A^\#,\end{aligned}$$

q.e.d.

CHAPTER III

Calculations with units

In this chapter we are going to use the descent machinery of I §3 in several settings: first for finite fields (an introductory example), next for ℓ-adic fields (where one recaptures a small part of local class field theory), and then for number fields. These calculations are basic for Chapter IV where $\mathbb{Z}_p$-extensions and Leopoldt's conjecture come into play.

We need a preliminary section which will allow us to bypass the problem that in number theory unit groups are usually only known "up to finite index" (cf. the standard form of Dirichlet's Unit Theorem). We fix a prime $\underline{p \neq 2}$.

§1 Results on twisted Galois modules

This section complements the results of §2 in Chap. II. Let $n \geq 1$, and Γ be a group with a fixed embedding ω: $\Gamma \longrightarrow (\mathbb{Z}/p^n)^*$. As in I 2.9, I 2.10, β will denote a generator of Γ, and $w \in \mathbb{Z}$ will be a lift of $\omega(\beta)$ such that $1 - w^m$ is divisible by p^n but not by p^{n+1} (with $m = |\Gamma|$). Finally, $\xi = \sum_{i=0}^{m-1} w^i \cdot \beta^{-i}$.

Theorem 1.1. *Let A be either a finitely generated $\mathbb{Z}[\Gamma]$-module, or a finitely generated $\mathbb{Z}_p[\Gamma]$-module, and suppose A has no p-torsion. Then for every Γ-submodule $B \subset A$ with finite index in A:*

$$\left|(A/p^n A)(-1)^\Gamma\right| = \left|(B/p^n B)(-1)^\Gamma\right| .$$

Proof. If A is a $\mathbb{Z}[\Gamma]$-module, then one has $A/p^n \cong A'/p^n$ for the $\mathbb{Z}_p[\Gamma]$-module $A' = \mathbb{Z}_p \otimes_{\mathbb{Z}} A$, so we need only treat the case that A and B are $\mathbb{Z}_p[\Gamma]$-modules. We then have by I 2.10: $\xi \cdot (1 - w^{-1}\beta) = qp^n$ with q a unit in $\mathbb{Z}_p$.

By II 2.1 we have $(A/p^n A)(-1)^\Gamma = (1-\beta) \cdot (A/p^n A)(-1) = \xi \cdot (A/p^n A)$. Since $\xi \cdot A$ contains $p^n A$, this can be rewritten as $\xi A/p^n A$. Thus we have to show:

$$[\xi A : p^n A] = [\xi B : p^n B].$$

The proof of this follows a familiar pattern. Consider first the four modules

$$\begin{array}{ccc} A & \supset & B \\ \cup & & \cup \\ \xi A & \supset & \xi B . \end{array}$$

Since ξ divides p^n in $\mathbf{Z}_p[\Gamma]$, and p acts injectively on A, also ξ acts injectively on A, and induces therefore an isomorphism $A/B \simeq \xi A/\xi B$. One obtains from this:

$$\begin{aligned} [A:\xi A][A:B] &= [A:\xi A][\xi A:\xi B] \\ &= [A:\xi B] \\ &= [A:B][B:\xi B], \end{aligned}$$

hence

$$[A:\xi A] = [B:\xi B]. \qquad (*)$$

With the same argument (ξ replaced by p^n) one gets

$$[A:p^n A] = [B:p^n B]. \qquad (**)$$

Dividing (**) by (*) yields the equality we want. (All indices in this calculation are finite. For this it suffices to see that $p^n B$, the smallest occuring module, has finite index in A, the largest occuring module, since $[A:B] < \infty$ and B is finitely generated over $\mathbf{Z}_p$.)

Remark. The theorem remains correct for Γ-coinvariants in the place of invariants, with a very similar proof.

Corollary 1.2. *Let A denote a finitely generated $\mathbf{Z}[\Gamma]$-module, or $\mathbf{Z}_p[\Gamma]$-module, A without p-torsion. Then the order of the group $(A/p^n)(-1)^\Gamma$ depends only on the $\mathbf{Q}[\Gamma]$-isomorphism class of $\mathbf{Q}\otimes_{\mathbf{Z}} A$ (or on the $\mathbf{Q}_p$-isomorphism class of $\mathbf{Q}_p\otimes_{\mathbf{Z}_p} A$, respectively). Furthermore, in the case of $\mathbf{Z}[\Gamma]$-modules, the order of $(A/p^n)(-1)^\Gamma$ depends only on the $\mathbb{R}[\Gamma]$-isomorphism classs of $\mathbb{R}\otimes_{\mathbf{Z}} A$.*

Proof. (For $\mathbf{Z}[\Gamma]$-modules; for $\mathbf{Z}_p[\Gamma]$-modules, the argument is even easier.) Since prime-to-p-torsion in A contributes nothing to A/p^n, and vanishes anyway when one tensors with $\mathbf{Q}$, we may suppose that A has no torsion. If B is another finitely generated $\mathbf{Z}[\Gamma]$-module without torsion, and $\mathbf{Q}\otimes A \simeq \mathbf{Q}\otimes B$, then, letting $M = \mathbf{Q}\otimes A$, we may consider both A and B as "lattices" in M, i.e. as finitely generated submodules containing a $\mathbf{Q}$-basis of M. Then $A\cap B$ is another lattice, hence it has finite index in A, and in B. Applying Thm. 1.1 twice, we get $|(A/p^n)(-1)^\Gamma| = |(B/p^n)(-1)^\Gamma|$ as claimed.

The statement concerning $\mathbb{R}\otimes A$ is a consequence, since it is well known from elementary representation theory that two $\mathbf{Q}[\Gamma]$-modules which become isomorphic on tensoring with $\mathbb{R}$ must already be isomorphic themselves.

Remark. Again, 1.2 remains correct for $(-)_\Gamma$ in the place of $(-)^\Gamma$.

Let us introduce a notation:

Definition. For any finitely generated module A over $\mathbb{Z}[\Gamma]$ or $\mathbb{Z}_p[\Gamma]$, let

$$\alpha(A) \;=\; \left|(A/p^nA)(-1)^\Gamma\right| \;\in\; \mathbb{N}.$$

Remark. Of course, $\alpha(A)$ is a power of p. From the proof of 1.1 one sees that $\alpha(A)$ also equals the order of $\xi A/p^nA$.

The following corollary is important in the sequel. Its first part follows from II 2.2, and its second part is just Cor. 1.2.

Corollary 1.3. *With the above notation:*

a) α is multiplicative on short exact sequences $0 \to A \to B \to C \to 0$ of $\mathbb{Z}[\Gamma]$-modules ($\mathbb{Z}_p[\Gamma]$-modules) such that C has no p-torsion: $\alpha(B) = \alpha(A)\cdot\alpha(C)$.

b) $\alpha(A)$ only depends on $\mathbb{Q}\otimes_{\mathbb{Z}} A$ (or $\mathbb{Q}_p\otimes_{\mathbb{Z}_p} A$, as the case may be) as long as A is chosen finitely generated and without p-torsion.

Thus, the general strategy to compute $\alpha(A)$ for arbitrary finitely generated A is as follows: Let $A' = A/\mathrm{tors}(A)$; then $\alpha(A) = \alpha(\mathrm{tors}(A))\cdot\alpha(A')$ by 1.3 a). Then replace A' by some $\mathbb{Z}[\Gamma]$-module A'' which is simple to handle and satisfies $\mathbb{Q}\otimes A' \cong \mathbb{Q}\otimes A''$ (and accordingly for $\mathbb{Z}_p[\Gamma]$-modules). For the second step, one might go through all simple $\mathbb{Q}[\Gamma]$-modules and pick $\mathbb{Z}$-lattices in them, but for our purpose the following result is more useful:

Lemma 1.4. *Let U be a subgroup of Γ, and $A = \mathbb{Z}[\Gamma/U]$ (or $\mathbb{Z}_p[\Gamma/U]$). For $i = 0, 1, \ldots, n$, let $Z^{(i)}$ be the kernel of $(\mathbb{Z}/p^n)^* \to (\mathbb{Z}/p^i)^*$, so that $Z^{(i)}$ is generated by the class of $1+p^i$ for $i \ge 1$. Then:*

$$\alpha(A) = 1 \qquad \textit{if } U \textit{ is not a } p\textit{-group};$$

$$\alpha(A) = p^i \qquad \textit{if } \omega(U) = Z^{(i)} \textit{ for some } i \in \{1, \ldots, n\}.$$

This is a complete list of cases, since $Z^{(1)}$ is the p-primary part of the cyclic group $(\mathbb{Z}/p^n)^*$, and all subgroups of $Z^{(1)}$ are of the form $Z^{(i)}$.

Proof. Let $x = \sum_{\bar\gamma\in\Gamma/U} a_{\bar\gamma}\cdot\bar\gamma \in (A/p^n)(-1)^\Gamma$ (with $a_{\bar\gamma} \in \mathbb{Z}/p^n$). Then $\omega(\beta)x = \beta x$ (recall β is a generator of Γ), i.e.

$$a_{\overline{\beta^{-1}\gamma}} = \omega(\beta)\cdot a_{\bar\gamma} \qquad \text{for } \gamma \in \Gamma/U.$$

Hence for all $\gamma \in \Gamma$: $a_{\overline{\gamma^{-1}}} = \omega(\gamma)\cdot a_1$. This forces that for $\delta \in U$ we have $a_1 = a_\delta = \omega(\delta)^{-1}a_1$, whence $a_1\cdot(1-\omega(\delta)) = 0$. If U is not a p-group, then there exists δ not in $\Gamma^{(1)}$, i.e. $\omega(\delta)$ not $\equiv$ to 1 (mod p), and the last equation forces $a_1 = 0 \in \mathbb{Z}/p^n$. Otherwise $\omega(U)$ is generated by $1+p^i$ $(1 \le i \le n)$, hence we obtain $a_1\cdot p^i = 0 \in \mathbb{Z}/p^n$. Conversely, one can als see that for any choice $a_1 \in p^{n-i}(\mathbb{Z}/p^n)$, the element

$$x = \sum_{\bar{\gamma}\in \Gamma/U} \omega(\gamma)^{-1} a_1 \cdot \bar{\gamma}$$

is in $(\mathbb{Z}/p^n[\Gamma/U])(-1)^\Gamma$, which proves the lemma since $p^{n-i}(\mathbb{Z}/p^n)$ has p^i elements.

§ 2 Finite fields and ℓ-adic fields

As an introduction, we first treat finite fields. In this case we do not even need the material of §1.

Let $\mathbb{F}$ be a finite field of characteric ℓ and cardinality $q = \ell^m$, and suppose $p \neq \ell$ is another prime. We know beforehand that $\mathrm{H}(\mathbb{F}, C_{p^n}) = \mathrm{NB}(\mathbb{F}, C_{p^n})$ must be cyclic of order p^n ($n \geq 1$ arbitrary), since the absolute Galois group $\Psi_{\mathbb{F}} = \mathrm{Aut}(\mathbb{F}^{sep}/\mathbb{F})$ is $\hat{\mathbb{Z}}$, the profinite completion of $\mathbb{Z}$, and therefore $\mathrm{H}(\mathbb{F}, C_{p^n}) \cong \mathrm{Hom}_{cont}(\hat{\mathbb{Z}}, C_{p^n}) \cong C_{p^n}$. Let us just demonstrate how this result also follows from Chap. I.

By I 2.7 and I 3.8, $\mathrm{NB}(\mathbb{F}, C_{p^n}) \cong (\mathbb{F}_n^*/p^n)(-1)^{\Gamma_n}$, where $\mathbb{F}_n = \mathbb{F}(\zeta_n) \subset \mathbb{F}^{sep}$, and $\Gamma_n = \mathrm{Aut}(\mathbb{F}_n/\mathbb{F})$. (NB. ζ_n is a primitive p^n-th root of 1.) Now $\mathbb{F}_n$ is the field with ℓ^r elements, where r is the smallest multiple of m such that $\mathbb{F}_{\ell^r}$ contains ζ_n, i.e. such that p^n divides $\ell^r - 1$. Hence r is the l.c.m. of m and the multiplicative order of ℓ modulo p^n. As one knows, Γ_n is generated by the Frobenius map φ: $x \longmapsto x^q$ ($x \in \mathbb{F}_n$). Here the usual embedding ω: $\Gamma_n \to (\mathbb{Z}/p^n)^*$ satisfies $\omega(\varphi) = q$, and this shows that Γ_n acts *trivially* on the twisted module $(\mathbb{F}_n^*/p^n)(-1)$. Hence $\mathrm{NB}(\mathbb{F}, C_{p^n}) \cong (\mathbb{F}_n^*/p^n)(-1)^{\Gamma_n} = \mathbb{F}_n^*/p^n$ as an abelian group. Since $\mathbb{F}_n^*$ is cyclic and its order is a multiple of p^n, the factor group $\mathbb{F}_n^*/p^n$ is indeed cyclic of order p^n, which we wanted to show.

Now we turn to ℓ-adic fields. We shall use the elementary theory of units and higher units in local fields. Let us point out two things: The results of this § will not be used in the sequel; and it seems that our methods are not so well suited for local function fields (which explains the restriction to the ℓ-adic case).

Let thus K be an ℓ-adic field, i.e. a finite extension of $\mathbb{Q}_\ell$, $\mathcal{O} = \mathcal{O}(K)$ the valuation ring of K (one frequently calls $\mathcal{O}$ the ring of integers in K), and $\mathfrak{p}$ the maximal ideal of $\mathcal{O}$. Exactly as in **0**§4 one shows: G-Galois extensions of $\mathcal{O}$ are in canonical bijection with *unramified* G-galois extensions of K (G finite group). From the theory of local fields one also has the standard result that for any finite abelian group G, there is a canonical bijection between the G-Galois extensions of $k = \mathcal{O}/\mathfrak{p}$, and the unramified G-Galois extensions of K. In our language this can be stated as follows: The map "tensor by $\mathcal{O}/\mathfrak{p}$ over $\mathcal{O}$" from $\mathrm{H}(\mathcal{O}, G)$ to $\mathrm{H}(k, G)$ is bijective.

Now let $G = C_{p^n}$. We want to calculate the order of $\mathrm{H}(K, C_{p^n})$. Let as above $K_n = K(\zeta_n) \subset K^{sep}$ (this is indeed the p^n-th cyclotomic extension of K in the sense of I §1), and $\Gamma_n = \mathrm{Aut}(K_n/K)$. By Cor. I 3.8, $|\mathrm{H}(K, C_{p^n})| = |\mathrm{NB}(K, C_{p^n})| = |(K_n^*/p^n)(-1)^{\Gamma_n}|$. The last term is just $\alpha(K_n^*)$ in the notation of §1 of this chapter. Let $D_n = K_n^*/\mathrm{tors}(K_n^*)$. Here clearly $\mathrm{tors}(K_n^*) = \mu_{p^\infty}(K_n) \times \mu'$, where μ' is the group of roots of unity in K_n whose order is prime to p. We need a lemma.

Lemma 2.1. *For any field K with $\mathrm{char}(K) \neq p$, such that $\mu_{p^\infty}(K(\zeta_1))$ is finite (in other words, $n_0(K) < \infty$), and $A = \mu_{p^\infty}(K(\zeta_n))$, we have*

$$|(A/p^n)(-1)^{\Gamma_n}| = p^n.$$

(NB. ζ_1 is a primitive p-th root of unity and ζ_n a primitive p^n-th root of unity.)

Proof. First case: $n \geq n_0(K)$. Then $A = \mu_{p^n}(K(\zeta_n))$ is cyclic of order p^n by I 1.7. and I 1.5. Then Γ_n operates trivially on $A(-1) = (A/p^n)(-1)$, by the very definition of the twist. Hence $(A/p^n)(-1)^{\Gamma_n}$ coincides with A, which suffices.

Second case: $n < n_0$. Considering the canonical surjection $\Gamma_{n_0} \to \Gamma_n$ and using $A = \mu_{p^\infty}(K(\zeta_{n_0}))$, we obtain that Γ_n still operates trivially on $(A/p^n)(-1)$. (The twisted Γ_n-module $A(-1)$ is not defined since $p^n A \neq 0$.) We get $|(A/p^n)(-1)^{\Gamma_n}| = |A/p^n| = p^n$, Q.E.D.

Given this lemma, we now may use Cor. 1.3 for our ℓ-adic field K to obtain $\alpha(K_n^*) = p^n \cdot \alpha(D_n)$. It is now well known that D_n sits in an exact sequence

$$1 \longrightarrow \mathrm{V}(K_n) \longrightarrow D_n \longrightarrow \mathbb{Z} \xrightarrow{v} 0,$$

where v is the valuation of K and $\mathrm{V}(K_n)$ is the factor group of the principal units $\mathrm{U}^1(K_n) = \{x \in \mathcal{O}(K_n) \mid x \equiv 1 \bmod \mathfrak{p}(K_n)\}$ modulo all p-power roots of unity. The sequence is Γ_n-equivariant, when we let Γ_n act trivially on $\mathbb{Z}$, i.e. $\mathbb{Z}$ can be written as the Γ_n-module $\mathbb{Z}[\Gamma_n/\Gamma_n]$. To calculate $\alpha(\mathbb{Z})$, we state another quick lemma.

Lemma 2.2. *Let Γ_n act trivially on $\mathbb{Z}$, and define m_K by the equation p^{m_K} = order of $\mu_{p^\infty}(K)$ (which is finite for any ℓ-adic field K). Then $\alpha(\mathbb{Z}) = p^{\min(n, m_K)}$.*

Proof. We apply 1.4 to $\mathbb{Z} = \mathbb{Z}[\Gamma_n/\Gamma_n]$, i.e. we let $U = \Gamma_n$. If $m_K = 0$, then Γ_n is not a p-group, hence $\alpha(\mathbb{Z}) = 1$. If $1 \leq m_K = m \leq n$, then $\zeta_1 \in K$, the number m_K equals the number $n_0(K)$, and Γ_n has order p^{n-m}, hence $\omega(\Gamma_n)$ must be the group $\mathbb{Z}^{(m)}$ (see 1.4), and $\alpha(\mathbb{Z}) = p^m$. If $1 \leq n < m_K$, then $K_n = K$, and Γ_n is trivial, hence $\alpha(\mathbb{Z}) = |\mathbb{Z}/p^n| = p^n$.

From the lemma and the preceding formulas we get, using 1.3 again:

$$\alpha(K_n^*) = p^n \cdot \alpha(\mathrm{V}(K_n)) \cdot p^{\min(n, m_K)}. \tag{1}$$

Now we let $\mathcal{O}_n = \mathcal{O}(K_n)$ and use the exponential function exp: $\ell^2\mathcal{O}_n \to 1+\ell^2\mathcal{O}_n$. One knows that the exp series converges on $\ell^2\mathcal{O}_n$ (for $\ell \neq 2$, exp converges already on $\ell\mathcal{O}_n$), and that it maps $\ell^2\mathcal{O}_n$ bijectively to $1+\ell^2\mathcal{O}_n$, the inverse function given by the ℓ-adic logarithm series. Thus the group $1+\ell^2\mathcal{O}_n$ is torsion-free, and has finite index in $\mathrm{U}^1(K_n) = 1+\pi\mathcal{O}_n$ (where π is a generator of $\mathfrak{p}(K_n)$). Therefore $1+\ell^2\mathcal{O}_n$ embeds into $\mathrm{V}(K_n)$ as a subgroup of finite index, and by 1.3 b) we get $\alpha(D_n) = \alpha(1+\ell^2\mathcal{O}_n)$, which equals $\alpha(\ell^2\mathcal{O}_n)$ by virtue of the exponential isomorphism (trivially, exp is Γ_n-equivariant). It is also clear that $\ell^2\mathcal{O}_n \cong \mathcal{O}_n$. Hence:

$$\alpha(\mathrm{V}(K_n)) = \alpha(\mathcal{O}_n). \tag{2}$$

We now have to distinguish two cases. The easy one is $\ell \neq p$. Then the (additive!) group $\mathcal{O}_n$ is p-divisible (since it is a $\mathbb{Z}_\ell$-module), hence by definition $\alpha(\mathcal{O}_n) = 1$. The interesting case is $\ell = p$. There we get $\mathbb{Q}_p \otimes \mathcal{O}_n = K_n$, and the $\mathbb{Q}_p[\Gamma_n]$-structure of K_n is quite well understood: Since K_n/K has a normal basis, we have an isomorphism $K_n \cong K[\Gamma_n]$ of $\mathbb{Q}_p[\Gamma_n]$-modules. Now $K[\Gamma_n] \cong \mathbb{Q}_p[\Gamma_n]^d$ with $d = [K:\mathbb{Q}_p]$. If we write $\mathbb{Q}_p[\Gamma_n]^d = \mathbb{Q}_p \otimes (\mathbb{Z}_p[\Gamma_n]^d)$ and invoke 1.3 b) again, we obtain $\alpha(\mathcal{O}_n) = \alpha(\mathbb{Z}_p[\Gamma_n]^d) = \alpha(\mathbb{Z}_p[\Gamma_n])^d$. Letting $U = 1$ in 1.4, we get

$$\begin{aligned} \alpha(\mathcal{O}_n) &= p^{nd} && \text{for } \ell = p, \\ \alpha(\mathcal{O}_n) &= 1 && \text{for } \ell \neq p. \end{aligned} \tag{3}$$

Putting together (1), (2), and (3), we finally obtain

Theorem 2.3. *For any ℓ-adic field K of degree d over $\mathbb{Q}_\ell$, and any $n \geq m_K$ (see 2.2) the following formulas hold:*

$$|\mathrm{H}(K, C_{p^n})| = p^{n+\min(n,m_K)} \cdot p^{nd} = p^{(d+1)n+\min(n,m_K)} \quad \textit{for } \ell = p,$$

$$|\mathrm{H}(K, C_{p^n})| = p^{n+\min(n,m_K)} \quad \textit{for } \ell \neq p.$$

(*Recall* $p^{m_K} = |\mu_{p^\infty}(K)|$.)

Any reader knowing some local class field theory will see how the statements in the last theorem follow from local CFT. Conversely, a small part of local CFT can be recaptured from 2.3. Consider the absolute Galois group $\Psi_K = \mathrm{Aut}(K^{sep}/K)$ of K, and let Ω be its maximal abelian pro-p-quotient, so Ω is the Galois group of the maximal abelian p-extension of K. Then $\mathrm{Hom}_{cont}(\Omega, C_{p^n}) = \mathrm{Hom}_{cont}(\Psi_K, C_{p^n}) \cong \mathrm{H}^1(K, C_{p^n})$. Letting $n = 1$ in 2.3, we obtain that $\mathrm{Hom}_{cont}(\Omega, C_p) = \mathrm{Hom}(\Omega/p, C_p)$ is finite, whence Ω is finitely generated, i.e. Ω is a product of a finite p-group F and e copies of $\mathbb{Z}_p$ for some $e \in \mathbb{N}$. Using 2.3 again, now for all n, one sees without difficulty that we must have $e = d+1$ (let $n \to \infty$) and F is cyclic of order m_K. Hence we "know" the group Ω. It goes without saying that there is much more to local class field theory than just the abstract knowledge of Ω.

§3 Number fields

The goal of this section is the calculation of the order of $\mathrm{NB}(\mathcal{O}_K[p^{-1}], C_{p^n})$ for any number field K, $n \geq 1$. As in this whole chapter, p is an odd prime. We shall write R for $\mathcal{O}_K[p^{-1}]$ throughout. For a description of $\mathrm{H}(R, C_{p^n})$ in terms of ramification, see 0§4.

The ring R is called *ring of p-integers* of K. It will be mentioned in passing how the results generalize to rings of S-integers, S any finite set of places which contains all places over p.

Let $d = [K:\mathbb{Q}]$, $d = r + 2s$, where r is the number of embeddings $K \longrightarrow \mathbb{R}$, and s is the number of conjugate pairs of nonreal embeddings $K \longrightarrow \mathbb{C}$. (These are customarily written r_1 and r_2; we need to economize on subscripts.) As earlier, S_n will denote the p^n-th cyclotomic extension of R. Recall our notation that ζ_n is a p^n-th primitive root of unity. The ring of p-integers in $K_n = K(\zeta_n)$ is $R[\zeta_n]$. (This is well-known, and may be checked locally as follows: The minimal polynomial f of ζ_n/K has no multiple roots modulo any prime $\mathfrak{p}$ of K not dividing p, hence $R_{\mathfrak{p}}[\zeta_n]/(\pi)$ is a finite product of fields (π a generator of $\mathfrak{p}R_{\mathfrak{p}}$). From this, one easily obtains that all localizations of $R_{\mathfrak{p}}[\zeta_n]$ are DVR's, hence $R[\zeta_n]$ is Dedekind, which suffices.) By 0§4, $R[\zeta_n]$ is Γ-Galois over R with $\Gamma = \mathrm{Aut}(K_n/K)$. By the uniqueness results of I §1, $\Gamma = \Gamma_n$ and we may identify S_n with $R[\zeta_n]$.

The first step is the theorem of Artin and Herbrand which gives us the Γ_n-module structure of $E(K_n)$, the unit group of $\mathcal{O}(K_n)$. As a preparation to its statement, let L/K be any Galois field extension with group G. Let $\beta_1, \ldots, \beta_r$ be the real embeddings of K, and $\beta_{r+1}, \ldots, \beta_{r+s}$ the nonreal embeddings of K modulo complex conjugation. For each β_i ($1 \leq i \leq r+s$) choose an embedding φ_i: $L \rightarrow \mathbb{C}$ extending β_i. Let j: $\mathbb{C} \rightarrow \mathbb{C}$ be the complex conjugation. Suppose that β_i is real but φ_i is nonreal. Then φ_i and $j\varphi_i$ agree on K, and have same image, hence there exists some $\tau_i \in G$ with $\varphi_i\tau_i = j\varphi_i$. One checks that $\tau_i \neq id$, $\tau_i^2 = id$. The aforementioned result now reads as follows:

Theorem 3.1. (Artin–Herbrand; see Artin (1931)) *Let τ_i be as above for all i such that β_i is real and φ_i is nonreal. Let $\tau_i = id$ for all other $i \in \{1,\ldots,r+s\}$. Then there exist units $\varepsilon_1, \ldots, \varepsilon_{r+s} \in E(L)$ such that the set $J = \{\gamma(\varepsilon_i) \mid \gamma \in G,\ 1 \leq i \leq r+s\}$ generates a subgroup of finite index in $E(L)$, and the relations between the elements of J are generated by the following ones:*

$$\tau_i(\varepsilon_i) = \varepsilon_i \quad \textit{for } i = 1, \ldots, r+s; \textit{ and}$$

$$\prod_{i=1,\ldots,r+s;\ \gamma\in G} \gamma(\varepsilon_i) = 1.$$

Let now $L = K_n$. Since $\zeta_n \in L$, *all* φ_i are nonreal. We claim: For each i such that β_i is real (and φ_i nonreal), the automorphism $\tau_i \in G = \Gamma_n$ defined above satisfies $\omega(\tau_i) = -1$. Proof: $\varphi_i(\zeta_n)$ is of necessity a p^n-th root of unity in $\mathbb{C}$, hence $j(\varphi_i(\zeta_n)) = \varphi_i(\zeta_n)^{-1}$. Thus $\varphi_i(\tau_i(\zeta_n)) = \varphi_i(\zeta_n)^{-1}$, which implies $\tau_i(\zeta_n) = \zeta_n^{-1}$, i.e. per definition: $\omega(\tau_i) = -1$. In the case that at least one β_i is real, i.e. $r \geq 1$, all τ_i $(i = 1,\ldots,r)$ are hence equal to one automorphism τ in Γ_n (which is characterized by $\omega(\tau) = -1$). If $r = 0$, let τ remain undefined. We may then reformulate Thm. 3.1 as follows: There exist $r+s$ units $\varepsilon_i \in E(K_n)$ such that the set $J = \{\gamma(\varepsilon_i) \mid 1 \leq i \leq r+s,\ \gamma \in \Gamma_n\}$ generates a subgroup of finite index in $E(K_n)$, and the relations are generated by two types of relations: first, $\tau(\varepsilon_i) = \varepsilon_i$ for $1 \leq i \leq r$, and second: the product of all $\gamma(\varepsilon_i)$ $(\gamma \in \Gamma_n,\ 1 \leq i \leq r+s)$ is unity.

For our purposes, the following rewording of 3.1 is useful:

Corollary 3.2. *With the above notation, define* $\Gamma_n' = \Gamma_n/\{id,\tau\}$. *(For* $r = 0$, Γ_n' *is not defined.) Then* $E(K_n)$ *contains a* Γ_n*-submodule of finite index isomorphic to the following* $\mathbb{Z}$*-free* Γ*-module:*

$$V^{r,s} = \left(\mathbb{Z}[\Gamma_n']^r \times \mathbb{Z}[\Gamma_n]^s\right)/\mathbb{Z}\cdot\left(\mathrm{N}_{\Gamma_n'}, \ldots, \mathrm{N}_{\Gamma_n'}, \mathrm{N}_{\Gamma_n}, \ldots, \mathrm{N}_{\Gamma_n}\right).$$

(Recall $\mathrm{N}_\Delta = \Sigma_{\delta\in\Delta}\,\delta \in \mathbb{Z}[\Delta]$ *for any finite group* δ.*).*

Proof. Let $e_1', \ldots, e_r'$ be the canonical $\mathbb{Z}[\Gamma_n']$-basis of $\mathbb{Z}[\Gamma_n']^r$ (ignore for $r = 0$), and let $e_{r+1}, \ldots, e_{r+s}$ be the canonical $\mathbb{Z}[\Gamma_n]$-basis of $\mathbb{Z}[\Gamma_n]^s$. By 3.1, we may define a Γ_n-linear map which sends e_i to ε_i $(1 \leq i \leq r+s)$, and this map defines an isomorphism $V^{r,s} \longrightarrow \mathbb{Z}\text{-span}\{\gamma(\varepsilon_i) \mid 1 \leq i \leq r+s,\ \gamma \in \Gamma_n\}$.

Observing that the $\mathbb{Z}[\Gamma_n]$-module $\mathbb{Z}(\mathrm{N}_{\Gamma_n'}, \ldots, \mathrm{N}_{\Gamma_n'}, \mathrm{N}_{\Gamma_n}, \ldots, \mathrm{N}_{\Gamma_n})$ is just $\mathbb{Z}$ with trivial Γ_n-action, we obtain the short exact sequence

$$0 \longrightarrow \mathbb{Z} \longrightarrow \mathbb{Z}[\Gamma_n']^r \times \mathbb{Z}[\Gamma_n]^s \longrightarrow V^{r,s} \longrightarrow 0.$$

Recall that m_K is defined by p^{m_K} = number of p-power roots of 1 in K. The torsion subgroup of the unit group $E(K_n)$ is $\mu(K_n)$, the group of all roots of unity in K_n.

Proposition 3.3. $\alpha(E(K_n)/\mu(K_n)) = p^{-\min(n,m_K)}\cdot p^{ns}$.

Proof. Since $V^{r,s}$ is torsion free, also $E(K_n)/\mu(K_n)$ contains a submodule of finite index isomorphic to $V^{r,s}$. Hence, by 1.2, we have $\alpha(E(K_n)/\mu(K_n)) = \alpha(V^{r,s})$. We use 1.3 and the above short exact sequence. Since $\alpha(\mathbb{Z}) = p^{\min(n,m_K)}$ by 2.2, we obtain $\alpha(V^{r,s}) = p^{-\min(n,m_K)}\cdot\alpha\left(\mathbb{Z}[\Gamma_n']^r \times \mathbb{Z}[\Gamma_n]^s\right)$. An easy argument for direct sums (or 1.3 if one likes) shows that $\alpha\left(\mathbb{Z}[\Gamma_n']^r \times \mathbb{Z}[\Gamma_n]^s\right) = \alpha(\mathbb{Z}[\Gamma_n'])^r\cdot\alpha(\mathbb{Z}[\Gamma_n])^s$. We are therefore done if we can show: $\alpha(\mathbb{Z}[\Gamma_n']) = 1$, and $\alpha(\mathbb{Z}[\Gamma_n]) = p^n$. The latter of these two formulas follows directly from 1.4, letting $U = \{id\}$. So does the former:

Suppose $r > 0$. By definition, $\Gamma_n' = \Gamma_n/U$ with $U = \{id, \tau\}$, and τ has order 2 (because of $\omega(\tau) = -1$). Hence U is not a p-group (because p is supposed to be odd!), and $\alpha(\Gamma_n/U) = 1$ by 1.4, q.e.d.

We are already approaching our aim of evaluating $|\mathrm{NB}(R, C_{p^n})|$. By I 3.8, this cardinality is equal to $|(S_n^*/p^n)(-1)^{\Gamma_n}|$, and this is $\alpha(S_n^*)$ by definition. It remains to connect up that quantity with $\alpha(E(K_n)/\mu(K_n))$. This is done in two steps.

Proposition 3.4. $\alpha(E(K_n)) \;=\; p^n\cdot\alpha(E(K_n)/\mu(K_n)) \;=\; p^{n-\min(n,m_K)}\cdot p^{ns}$.

Proof. The second equality is just Prop. 3.3. For the first, we look at the sequence $1 \to \mu(K_n) \to E(K_n) \to E(K_n)/\mu(K_n) \to 1$. By 1.3 it suffices to see that $\alpha(\mu(K_n)) = p^n$. Now certainly $\alpha(\mu(K_n)) = \alpha(\mu_{p^\infty}(K_n))$ (in the definition of α, one goes mod p^n first), and the desired equality follows from 2.1.

For the next (and last) auxiliary result in this §, we introduce an abbreviation: if v is a place of K, we write m_v instead of m_{K_v}. We then may state:

Proposition 3.5. $\alpha(S_n^*) \;=\; \alpha(E(K_n))\cdot\prod_{v|p} p^{\min(n,m_v)}$.

Proof. Since S_n is the ring of p-integers of K_n, the unit group S_n^* is the group of p-units in K_n, i.e. the group of all $x \in K_n^*$ whose valuation is zero at all v not dividing p. Hence we get another exact Γ_n-equivariant sequence

$$1 \longrightarrow E(K_n) \longrightarrow S_n^* \xrightarrow{W} \prod_{w|p} \mathbb{Z} = P \qquad (w \text{ denoting places of } K_n),$$

with $W(x) = (w(x))_{w|p}$, and Γ_n acting on P via permutation of the factors $\mathbb{Z}$ (of course, Γ_n permutes the places w of K_n over p). Now the cokernel of W injects into the class group of K_n (send the w-th canonical basis vector of P to the prime ideal $\mathfrak{p}_w$ corresponding to w), and hence is finite. This yields by 1.2 that $\alpha(\mathrm{Im}(W)) = \alpha(P)$, hence $\alpha(S_n^*) = \alpha(E(K_n))\cdot\alpha(P)$. It remains to evaluate $\alpha(P)$.

By grouping the w dividing p into Γ_n-orbits, we obtain a Γ_n-isomorphism

$$P \;=\; \prod_{w|p} \mathbb{Z} \cong \prod_{v|p} \mathbb{Z}[\Gamma_n/\Gamma_v] \qquad (v \text{ denotes now places of } K),$$

where Γ_v is the stabilizer of some, and hence all, w dividing v under Γ_n. Γ_v is commonly called the decomposition group of v. We have used that Γ_n is abelian, and that all w dividing a given v are conjugate under Γ_n. One knows that $\Gamma_v \cong \mathrm{Aut}(K_v(\zeta_n)/K_v)$. We distinguish three cases:

If ζ_1 is not in K_v, then Γ_v is not a p-group, and by 1.4 we have
$\alpha(\mathbb{Z}[\Gamma_n/\Gamma_v]) = 1$.

If $\zeta_m \in K_v$, ζ_{m+1} not in K_v, and $m \le n$, then one sees easily that ζ_n has degree

p^{n-m} over K_v, hence $|\Gamma_v| = p^{n-m}$, and we must have $\omega(\Gamma_v) = Z^{(m)}$ in the notation of 1.4. By that lemma, $\alpha(\mathbb{Z}[\Gamma_n/\Gamma_v]) = p^m$.

If ζ_n is already in K_v, then Γ_v is trivial and $\alpha(\mathbb{Z}[\Gamma_n/\Gamma_v]) = p^n$ by 1.4. Thus in every case we get $\alpha(\mathbb{Z}[\Gamma_n/\Gamma_v]) = p^{\min(n,m_v)}$, which finishes the proof.

By combining 3.4 and 3.5 and recalling $|\mathrm{NB}(R, C_{p^n})| = \alpha(S_n^*)$ we finally obtain:

Theorem 3.6. *Let p be an odd prime. For any number field K with r real and 2s nonreal embeddings, we have for the ring $R = \mathcal{O}_K[p^{-1}]$ of p-integers of K, and $n \geq 1$, the formula*

$$|\mathrm{NB}(R, C_{p^n})| = p^{-\min(n,m_K)} . \prod_{v|p} p^{\min(n,m_v)} . p^{n(s+1)}.$$

(For the definition of m_K and m_v see above.)

Example: Let $K = \mathbb{Q}$. Here $s = 0$, $r = 1$, $m_K = 0$. There is only one v dividing p, namely $v = v_p$ itself, and m_v is also zero. Hence the formula simplifies to $|\mathrm{NB}(R, C_{p^n})| = p^n$ for all $n \geq 1$. Since the cyclotomic p-extensions A_n of R generate a subgroup of order p^n in $\mathrm{NB}(R, C_{p^n})$, we obtain: Every abelian p-extension of R with normal basis is cyclotomic. This is a piece of the famous Kronecker-Weber theorem.

We may also consider larger rings than R. Let Σ be any finite set of finite places of K such that all v dividing p are in Σ. Let R_Σ be the ring of Σ-integers in K. One has the following result, whose proof is exactly analogous to 3.6 (it will not be used in the sequel):

Theorem 3.7. *With the notation of* 3.6, *one has*

$$|\mathrm{NB}(R_\Sigma, C_{p^n})| = p^{-\min(n,m_K)} . \prod_{v\in\Sigma} p^{\min(n,m_v)} . p^{n(s+1)}.$$

In the next chapter, we will be concerned with the asymptotic behaviour of $|\mathrm{NB}(R, C_{p^n})|$ and $|\mathrm{H}(R, C_{p^n})|$ for $n \to \infty$. Let us agree on the following Landau notation: if φ and ψ are functions $\mathbb{N} \longrightarrow \mathbb{R}^+$, we shall write

$$\varphi(n) = O(1)\cdot\psi(n)$$

if there are positive constants C, $D \in \mathbb{R}$ such that $C\varphi(n) \leq \psi(n) \leq D\varphi(n)$ for all $n \in \mathbb{N}$. From 3.6 we immediately infer:

Corollary 3.8. $|\mathrm{NB}(R, C_{p^n})| = O(1)\cdot p^{n(s+1)}$.

CHAPTER IV

Cyclic p-extensions and $\mathbb{Z}_p$-extensions of number fields

§1 C_{p^n}-extensions and ramification

The purpose of this section is to collect some facts from class field theory and to derive some consequences concerning the groups $\mathrm{H}(R, C_{p^n})$ (with p a fixed prime, $n \geq 1$, and R = ring of p-integers in a number field K). This should be seen in connection with III §3 where we determined the size of the group $\mathrm{NB}(R, C_{p^n})$. Our standard reference is the appendix of the book of Washington (1982).

Let thus K be a number field. For each place v of K, let K_v be the completion of K at v as usual, and

$$U_v = \text{the unit group of the valuation ring } \mathcal{O}(K_v) \text{ for } v \text{ nonarchimedean;}$$
$$U_v = K_v^* \text{ for } v \text{ archimedean.}$$

Let $\mathrm{I}_K = \{(x_v) \mid x_v \in K_v \text{ for each place } v;\ x_v \in U_v \text{ for almost all } v\}$ be the idele group of K. The group K^* embeds into I_K diagonally, and I_K has a topology in which it is locally compact, and K^* is discrete. The factor group $\mathrm{C}_K = \mathrm{I}_K / K^*$ is called the idele class group of K. Now the main theorem of global class field theory reads in our setting:

$$\mathrm{H}(K, G) \cong \mathrm{Hom}_{cont}(\mathrm{C}_K, G)$$

functorially in the finite group G. (This follows from **0**§3 since C_K mod connected component of 1 is canonically (via the Artin map) isomorphic to Ψ_K^{ab}, the Galois group of the maximal abelian extension of K.) Moreover one has precise control over ramification: If $L \in \mathrm{H}(K, G)$ belongs to $f_L \colon \mathrm{C}_K \longrightarrow G$, then L is unramified at v iff f_L is trivial on the image of U_v in C_K. [If $L \in \mathrm{H}(K,G)$ is not a field, i.e. if L comes from a field $L_0 \in \mathrm{H}(R, G_0)$, $G_0 \subset G$, then we say L is unramified at v if L_0 is.] Putting $R = \mathcal{O}(K)[p^{-1}]$ as always, we obtain:

Lemma 1.1. $\mathrm{H}(R, G) \cong \{L \in \mathrm{H}(K, G) \mid L$ *unramified at all finite places* v *not over* $p\} = \mathrm{Hom}_{cont}(\mathrm{C}_K/\mathrm{U}_0, G)$ *with* U_0 = *image of the product of all* U_v *(*v *finite, not over* p*) in* C_K. *Moreover it amounts to the same when we replace* U_0 *by* $\mathrm{U} = \mathrm{U}_0 \cdot$*image of* V,

$$V \stackrel{def}{=} \prod_{v\ real} \mathbb{R}^+ \times \prod_{v\ complex} \mathbb{C}^*,$$

because the image of V in C_K lies in the connected component of $1 \in C_K$.

Lemma 1.2. C_K/U *is a profinite group, and isomorphic to* $\mathbb{Z}_p^{c(K)} \times$ *(finite group) for a certain* $c(K) \in \mathbb{N}$.

Proof. For v finite, one has $K_v^*/U_v \simeq \mathbb{Z}$ via v. Hence

$$C_K/U \simeq \left(\prod_{v|p} K_v^* \times \prod_{v\nmid p, v\nmid\infty} \mathbb{Z} \times \prod_{v\ real} \{1,-1\}\right)\Big/\overline{\mathrm{Im}(K^*)}.$$

Let $D =$ image of $\prod_{v\ real}\{1,-1\}$ in C_K. Then $C_K/U{\cdot}D \simeq \left(\prod_{v|p} K_v^* \times \prod_{v\nmid p} \mathbb{Z}\right)/\mathrm{Im}(K^*)$. There is the canonical map π: $C_K/U{\cdot}D \longrightarrow \mathrm{Cl}(K)$. For $v|p$, it maps U_v to 1 and a parameter of K_v^* to $\mathfrak{p}_v$. For v not over p, it maps $1 \in \mathbb{Z}$ to $\mathfrak{p}_v$. We have:

$$\mathrm{Ker}(\pi) \simeq \left(\prod_{v|p} U_v\right)\Big/\overline{\mathrm{Im}(E_K)} \qquad (E_K = \mathcal{O}(K)^*).$$

This may be checked by defining a map f going from left to right by putting in 1's at all places v not over p, and verifying it is an isomorphism. Now $\prod_{v|p} U_v$ is profinite and of the form $\mathbb{Z}_p^m \times$(finite) (actually $m = [K:\mathbb{Q}]$). Hence also $\mathrm{Ker}(\pi)$ is a product of finitely many copies of $\mathbb{Z}_p$ and a finite group. Since the three groups $\mathrm{Ker}(\pi)$, C_K/UD, and C_K/U only differ by finite groups, the lemma follows.

Remark. By Dirichlet's Unit Theorem, the $\mathbb{Z}$-rank of E_K is $r+s-1$, where, as before, r is the number of real embeddings of K, and $2s$ the number of nonreal embeddings. This implies at once that $\overline{\mathrm{Im}(E_K)} \subset \prod_{v|p} U_v$ is of the form (finite)$\times\mathbb{Z}_p^{r+s-1-\delta}$ with $\delta \geq 0$. Leopoldt's Conjecture states that δ is zero for all number fields K. Given the fact that $\prod_{v|p} U_v =$ (finite)$\times\mathbb{Z}_p^{r+2s}$, we conclude that $\mathrm{Ker}(\pi)$ (or equivalently, C_K/U) is of the form (finite)$\times\mathbb{Z}_p^{s+1+\delta}$. An equivalent formulation of Leopoldt's conjecture is therefore: The number $c(K)$ in Lemma 1.2 is precisely $s+1+\delta$ (and not larger). Cf. Washington (1982) [§5.5].

Corollary 1.3. *The asymptotic growth of* $|\mathrm{H}(R,C_{p^n})|$ *is given by*

$$|\mathrm{H}(R,C_{p^n})| = O(1)\cdot p^{n\cdot c(K)}. \qquad \textit{(The } O \textit{ notation is defined in III §3.)}$$

Proof. Write $C_K/U \simeq \mathbb{Z}_p^{c(K)} \times F$, F finite, as in 1.2. By the initial remarks of this §, we have $\mathrm{H}(R,C_{p^n}) \simeq \mathrm{Hom}_{cont}(C_K/U,C_{p^n}) \simeq \mathrm{Hom}(\mathbb{Z}_p^{c(K)} \times F, C_{p^n})$, and the cardinality of the latter Hom lies between $p^{nc(K)}$ and $|F|\cdot p^{nc(K)}$, q.e.d.

Corollary 1.4. *Leopoldt's Conjecture holds for K and p iff* $|\mathrm{P}(R,C_{p^n})|$ *is bounded for* $n \to \infty$. *(Recall* $\mathrm{P}(R,C_{p^n}) = \mathrm{H}(R,C_{p^n})/\mathrm{NB}(R,C_{p^n})$.)

Proof. It was shown in III 3.8 that $|\mathrm{NB}(R,C_{p^n})| = O(1)\cdot p^{n(s+1)}$. From 1.3 we obtain by comparison that $|\mathrm{P}(R,C_{p^n})| = O(1)\cdot p^{n(c(K)-s-1)}$. This first of all reproves the fact $c(K) \geq s+1$; second, it visibly shows that one has equality $c(K) = s+1$ iff the quantity $|P(R, C_{p^n})|$ is bounded for $n \to \infty$. By the above remark, Leopoldt's conjecture is equivalent to $c(K) = s+1$.

§2 $\mathbb{Z}_p$-extensions

For a general introduction to $\mathbb{Z}_p$-extensions we refer to **0** §8. Here we shall be occupied with number fields and rings of p-integers. Let K and R be as in §1.

From Lemma 1.1 we get: $H(R, \mathbb{Z}_p) = \varprojlim H(R, C_{p^n}) \cong \varprojlim \mathrm{Hom}_{cont}(C_K/U, C_{p^n}) = \mathrm{Hom}_{cont}(C_K/U, \mathbb{Z}_p)$. With the help of Lemma 1.2 this gives:

Lemma 2.1. $H(R, \mathbb{Z}_p)$ *is finitely generated free of rank* $c(K)$ *over* $\mathbb{Z}_p$.

Remark. By Prop. 13.2 of Washington (1982), $\mathbb{Z}_p$-extensions of K are unramified in all finite v not dividing p, hence $H(R, \mathbb{Z}_p) = H(K, \mathbb{Z}_p)$. (The analog for C_{p^n}-extensions does not hold.) Thus, Lemma 2.1 says that K has exactly $c(K)$ independent $\mathbb{Z}_p$-extensions. One gets another equivalent condition for Leopoldt's conjecture to hold (for K and p): The number of independent $\mathbb{Z}_p$-extensions of K must be $s+1$. In particular, if K is totally real, the cyclotomic $\mathbb{Z}_p$-extension (= the union of all $A_n(K)$, see I §1) must, in a sense, be the only $\mathbb{Z}_p$-extension of K.

The main result of this section is that the subgroup $NB(R, \mathbb{Z}_p)$ has the $\mathbb{Z}_p$-rank $s+1$ which is predicted by Leopoldt's conjecture for $H(R, \mathbb{Z}_p)$. As a corollary, that conjecture holds iff $NB(R, \mathbb{Z}_p)$ has finite index in $H(R, \mathbb{Z}_p)$.

Theorem 2.2. *For all number fields* K *and all* odd *primes* p *there is an isomorphism*

$$NB(R, \mathbb{Z}_p) \cong \mathbb{Z}_p^{s+1}. \qquad (2s = \textit{number of nonreal embeddings of } K)$$

Remark. This result was proved for CM-fields K by Kersten and Michaliček (1988), in the general case by the author (1988,1991). (For a definition of CM-fields, see below.) A shorter proof which uses more number-theoretical apparatus was recently given by Fleckinger and Nguyen Quang Do (1991). For a quick argument in the totally real case, see Greither (1991) [Thm. 4.4]. It is not too hard to deduce the full CM case from this.

For the *proof* of **2.2** we need some preparation.

We fix a generator σ_n of C_{p^n} for all n. There are canonical projections ψ_n: $C_{p^{n+1}} \to C_{p^n}$ given by $\sigma_{n+1} \mapsto \sigma_n$, and canonical injections ι_n: $C_{p^n} \to C_{p^{n+1}}$ given by

$\sigma_n \mapsto \sigma_{n+1}^p$. For $m > n$, let $\iota_{n,m}$ be the composition $\iota_{m-1}\cdot\ldots\cdot\iota_n$: $C_{p^n} \to C_{p^m}$. Let $A_n = \mathrm{NB}(R, C_{p^n})$, $B_n = \mathrm{H}(R, C_{p^n})$. Then we have maps ι_n^*: $A_n \to A_{n+1}$, ι_n^*: $B_n \to B_{n+1}$. If we identify B_n with $\mathrm{Hom}_{cont}(\Psi_K, C_{p^n})$, then ι_n^* is just $\mathrm{Hom}(\Psi_K, \iota_n)$. Hence all ι_n^* are injective (this is, by the way, already proved in **0** §3). Let $B_\infty = \lim_\to B_n \cong \mathrm{Hom}_{cont}(\Psi_K, \mathbb{Q}_p/\mathbb{Z}_p)$. (We have used the identification $\lim_\to C_{p^n} = \mathbb{Q}_p/\mathbb{Z}_p$, $\sigma_n \mapsto p^{-n} + \mathbb{Z}_p$.) Similarly, $A_\infty = \lim_\to A_n \subset B_\infty$.

Now $B_n = \mathrm{Hom}_{cont}(W, C_{p^n})$ with $W = \mathrm{C}_K/\mathrm{U}$ (see 1.1) isomorphic to $\mathbb{Z}_p^{c(K)} \times F$, $F = \mathrm{tors}(W)$ a finite abelian group. Since (under the above identification) C_{p^n} is equal to $(\mathbb{Q}_p/\mathbb{Z}_p)[p^n]$, this implies

$$B_n \cong B_\infty[p^n] \text{ canonically for all } n \geq 1.$$

(For any abelian group M, we let $M[p^n] = \{x \in M | p^n x = 0\}$.) Furthermore, $B_\infty \cong \mathrm{Hom}(W, \mathbb{Q}_p/\mathbb{Z}_p) \cong (\mathbb{Q}_p/\mathbb{Z}_p)^{c(K)} \times$(finite group). It is fairly easy to see that any subgroup $M \subset B_\infty$ is isomorphic to $(\mathbb{Q}_p/\mathbb{Z}_p)^i \times$(finite group) for some $i \leq c(K)$. (Use a direct argument, or Pontryagin duality and the theory of finitely generated modules over the PID $\mathbb{Z}_p$.) Applying this to A_∞, we see $A_\infty = A' \oplus A''$ with A'' finite and $A' \cong (\mathbb{Q}_p/\mathbb{Z}_p)^i$. Since A' is then divisible, A' must already lie in $(\mathbb{Q}_p/\mathbb{Z}_p)^{c(K)}$, the divisible part of B_∞.

We certainly have $A_n \subset A_\infty[p^n]$ since A_n is killed by p^n. The central point of the argument is that this inclusion is "not too far from an equality".

Theorem 2.3. *The index e_n of A_n in $A_\infty[p^n]$ is bounded independently of n.*

Let us first show how Thm. 2.2 follows from 2.3. Thm. 2.3 gives at once that $|A_n| = O(1)\cdot|A_\infty[p^n]|$, and this equals $O(1)\cdot p^{ni}$ by the above description of A_∞. From III 3.8 we infer that $i = s+1$. It also follows from 2.3 that there exists a constant $c \in \mathbb{N}$ such that

$$p^c \cdot A'[p^n] \subset A_n \subset A'[p^n] \oplus A'' \quad \text{for all } n. \qquad (*)$$

Now we have for every $x \in B_{n+1}$ that $\pi_n^* x = p\cdot x$ under our identifications: $B_n = \mathrm{Hom}(W, C_{p^n})$ and $B_{n+1} = \mathrm{Hom}(W, C_{p^{n+1}})$ are identified with their images in $\mathrm{Hom}(W, \mathbb{Q}_p/\mathbb{Z}_p)$, and π_n is then just multiplication by p: $\langle p^{-n-1} + \mathbb{Z}_p\rangle \to \langle p^{-n} + \mathbb{Z}_p\rangle$. Hence also on A_{n+1} the map π_n acts as multiplication by p. We pass to the inverse limit in (*) over n and obtain:

$$\lim_\leftarrow p^c \cdot A'[p^n] \subset \lim_\leftarrow A_n \subset \lim_\leftarrow A'[p^n] \oplus \lim_\leftarrow A''.$$

Since all π_n just act via multiplication by p and $A' \cong (\mathbb{Q}_p/\mathbb{Z}_p)^{s+1}$, both $\lim_\leftarrow p^c \cdot A'[p^n]$ and $\lim_\leftarrow A'[p^n]$ are isomorphic to $\mathbb{Z}_p^{s+1}$. Moreover, $\mathrm{NB}(R, \mathbb{Z}_p) = \lim_\leftarrow A_n$ has no torsion (since it sits in $\mathrm{Hom}(W, \mathbb{Z}_p)$). From this one sees that $\mathrm{NB}(R, \mathbb{Z}_p)$ must be isomorphic to $\mathbb{Z}_p^{s+1}$.

For the *proof* of Thm. 2.3, we first quote a result of Iwasawa theory: Consider all base extension maps $\mathrm{Pic}(S_n)[p^n] \longrightarrow \mathrm{Pic}(S_m)[p^m]$ for $m \geq n$. Then by Iwasawa (1973) [p. 259], there is a constant c such that the kernel of all these maps has order at most p^c. We shall show that $e_n \leq p^c$ for all n. Recall that e_n = index of A_n in $A_\infty[p^n]$; by (1) above we also have $A_\infty[p^n] = A_\infty \cap B_n$. Since $A_\infty \cap B_n$ is the union of all $A_m \cap B_n$ $(m \geq n)$, it suffices to show $[A_m \cap B_n : A_n] \leq p^c$ for all $m \geq n$.

Consider for $n \leq m$ the commutative diagram

$$\begin{array}{ccccccccc} 0 & \longrightarrow & A_n & \longrightarrow & B_n & \longrightarrow & \mathrm{P}(R, C_{p^n}) & \longrightarrow & 0 \\ & & \downarrow \iota^*_{n,m} & & \downarrow \iota^*_{n,m} & & \downarrow \iota^*_{n,m} & & \\ 0 & \longrightarrow & A_m & \longrightarrow & B_m & \longrightarrow & \mathrm{P}(R, C_{p^m}) & \longrightarrow & 0 . \end{array}$$

We consider the two left-hand vertical maps as inclusions, as above. Then $A_m \cap B_n / A_n \cong \mathrm{Ker}(A_m/A_n \longrightarrow B_m/B_n)$, and this is by the snake lemma isomorphic to the kernel of the rightmost vertical map. Now $\mathrm{P}(R, C_{p^n})$ embeds by I 3.7 and 0 6.5 into $\mathrm{Pic}(S_n)$ via the map $\pi_n j_n$ (j_n the base change from R to S_n, and π_n the isomorphism from Kummer theory). We are therefore done if we have shown the following lemma:

Lemma 2.4. *The diagram*

$$\begin{array}{ccc} \mathrm{P}(R, C_{p^n}) & \xrightarrow{\pi_n j_n} & \mathrm{Pic}(S_n) \\ \iota^*_{n,m} \downarrow & & \downarrow S_m \otimes_{S_n} - \\ \mathrm{P}(R, C_{p^m}) & \xrightarrow{\pi_m j_m} & \mathrm{Pic}(S_m) \end{array}$$

is commutative, if $\zeta_n \in S_n$ and $\zeta_m \in S_m$ are chosen "correctly", i.e. such that $\zeta_m^{p^{m-n}} = \zeta_n$.

Proof. First of all, we may assume $m = n+1$. Second, since $\mathrm{H}(R,-)$ surjects onto $\mathrm{P}(R,-)$, we may replace P by H in both occurences. Abbreviate ζ_n to ζ and ζ_{n+1} to ζ', and similarly σ_n to σ and σ_{n+1} to σ'. NB. $\sigma'^p = \sigma$. Recall that π_n of some $B \in \mathrm{H}(S_n, C_{p^n})$ is (the class of) the invertible S_n-module $B^{(\zeta-\sigma)} = \{x \in B \mid \sigma x = \zeta x\}$. Now let us pick $A \in \mathrm{H}(R, C_{p^n})$ and chase it both ways. Let $B = S_n \otimes_R A$. Then via the upper right hand corner, A goes to $S_{n+1} \otimes_{S_n} (B^{\zeta-\sigma}) = (S_{n+1} \otimes_{S_n} B)^{\zeta-\sigma} = C^{\zeta-\sigma}$ with $C = (S_{n+1} \otimes_{S_n} B)$. (For the first equality, we used the flatness of S_{n+1} over S_n.)

Let us now calculate $S_{n+1} \otimes_R \iota_n^* A$:

$$S_{n+1} \otimes_R \iota_n^* A \quad \cong \quad \iota_n^* C \qquad \text{(since } \iota_n^* \text{ commutes with base change)},$$

and using the definition one checks

$$\iota_n^* C \quad \cong \quad C \times \ldots \times C \qquad \text{(factors numbered from 0 to } p-1),$$

with Galois action given by $\sigma'(c_0,c_1,\ldots,c_{p-1}) = (\sigma(c_{p-1}),c_0,c_1,\ldots,c_{p-2})$. We now calculate $\pi_{n+1}j_{n+1}(A) = \pi_{n+1}(\iota_n{}^*C) = (\iota_n{}^*C)^{\zeta'-\sigma'}$:

$$(c_0,c_1,\ldots,c_{p-1}) \in (\iota_n{}^*C)^{\zeta'-\sigma'} \iff$$

$$\iff (\sigma(c_{p-1}),c_0,c_1,\ldots,c_{p-2}) = (\zeta'c_0,\zeta'c_1,\ldots,\zeta'c_{p-1})$$

$$\iff c_1 = \zeta'^{-1}c_0, \ldots, c_{p-1} = \zeta'^{-1}c_{p-2}, \text{ and } c_0 = \zeta'^{-1}\sigma(c_{p-1})$$

$$\iff (c_0,c_1,\ldots,c_{p-1}) = (c_0,\zeta'^{-1}c_0,\ldots,\zeta'^{-p+1}c_0), \text{ and also}$$

$$c_0 = \zeta'^{-1}\cdot\sigma(\zeta'^{-p+1}c_0) = \zeta'^{-p}\cdot\sigma(c_0).$$

Now $\zeta'^p = \zeta$, so the last equation is equivalent to $\sigma(c_0) = \zeta c_0$. Hence the association $(c_0,c_1,\ldots,c_{p-1}) \longmapsto c_0$ defines an isomorphism $(\iota_n{}^*C)^{\zeta'-\sigma'} \longrightarrow C^{\zeta-\sigma}$. This shows that the diagram commutes. Lemma 2.4, and hence also Thm 2.3 and 2.2, are now proved.

Remark. There is a result similar to Lemma 2.4 concerning the projections ψ_n: $C_{p^{n+1}} \to C_{p^n}$, to wit: the following diagram commutes.

$$\begin{array}{ccc}
\mathrm{H}(R,C_{p^{n+1}}) & \longrightarrow & \mathrm{Pic}(S_{n+1}) \\
\pi_n^* \big\downarrow & & \big\downarrow \text{Norm from } K_{n+1} \text{ to } K_n \\
\mathrm{H}(R,C_{p^n}) & \longrightarrow & \mathrm{Pic}(S_n)\ .
\end{array}$$

For a proof, see Greither (1991) [II 5.5].

Corollary 2.5. *Leopoldt's Conjecture holds for K and p iff* $\mathrm{NB}(R,\mathbb{Z}_p)$ *has finite index in* $\mathrm{H}(R,C_{p^n})$.

Proof. Since the Leopoldt conjecture is equivalent to "$s+1 = c(K)$" (see §1), this is clear from Thm. 2.2 and Lemma 2.1.

For examples and applications, see §4 and §5.

§ 3 The asymptotic order of $P(R, C_{p^n})$

As always, let K be a number field with $2s$ nonreal embeddings, $p \neq 2$, $R = \mathcal{O}_K[p^{-1}]$. In III 3.6 we have shown that

$$|\mathrm{NB}(R, C_{p^n})| = c \cdot p^{n(s+1)} \quad \text{for } n \text{ large,}$$

with $c = p^{-m_K} \cdot \prod_{v|p} p^{m_v}$. (Actually, $n \geq \max_{v|p} m_v$ already suffices.)

On the other hand, we can write the group $W = C_K/U$ in the form $W \cong \mathbb{Z}_p^{c(K)} \times F$ with F finite, as in §1. We then get from $\mathrm{H}(R, C_{p^n}) = \mathrm{Hom}(W, C_{p^n})$ that

$$|\mathrm{H}(R, C_{p^n})| = |F|_p \cdot p^{nc(K)} \quad \text{for } n \text{ large,}$$

where $|F|_p$ is the p-part of the order of F. These two statements combined yield:

Theorem 3.1. *For $n \to \infty$, the order of $\mathrm{P}(R, C_{p^n})$ either goes to infinity (precisely if $s+1 < c(K)$, i.e. Leopoldt's Conjecture fails for K and p), or it is eventually constant with value $|F|_p \cdot p^{-m_K} \cdot \prod_{v|p} p^{m_v}$.*

This motivates the following notation:

Definition. The number q_K (or $q_{K,p}$ if p is not given by the context) is ∞ if $s+1 < c(K)$; $q_K = |F|_p \cdot p^{-m_K} \cdot \prod_{v|p} p^{m_v}$ as in 3.1 otherwise. Equivalently:

$$q_K = \lim_{n\to\infty} |\mathrm{P}(R, C_{p^n})| = \lim_{n\to\infty} \left(|\mathrm{H}(R, C_{p^n})| / |\mathrm{NB}(R, C_{p^n})| \right).$$

Note that we don't have a single example with $q_K = \infty$. On the other hand, it is not so easy in general to compute q_K, the main problem being the factor $|F|_p$. In some way q_K behaves like a class number. We shall try in this section to give some information on q_K in two particular cases: a) K is totally real, and b) K is a CM field (see below). In the latter case, we shall be interested only in the "minus part" of q_K, since the "plus part" will turn out to be just q_{K^+}.

Let us therefore assume now that K is totally real. We need a few more notations: $h = h_K$ = class number of K, $\Delta = \Delta_K$ the discriminant of $K/\mathbb{Q}$, $K_1 = K(\zeta_1)$ (remember ζ_1 is a primitive p-th root of unity), and $R_p = R_{K,p} \in \mathbb{C}_p$ the p-adic regulator of K. For the definition of R_p we have to refer to Washington (1982) [§5.5]. Let $\delta_K = \infty$ for $s+1 < c(K)$, and δ_K defined by $p^{\delta_K} = |F|_p$ otherwise. Slightly abusing notation, we can then write $q_K = p^{\delta_K} \cdot p^{-m_K} \cdot \prod_{v|p} p^{m_v}$. There is the following result:

Theorem 3.2. *Let as above K be totally real. Then $\delta_K - m_{K_1}$ is the p-valuation of*

$$h \cdot R_p \cdot \Delta^{-1/2} \cdot \prod_{v|p} (1 - \mathrm{N}v^{-1}) \quad \in \quad \mathbb{C}_p.$$

Proof. If Leopoldt's conjecture should fail for K and p, then $R_p = R_{K,p} = 0$ by loc.cit., and $\delta_K = \infty$, so the theorem holds. Suppose therefore $s+1 = c(K)$, hence $c(K) = 1$ since $s = 0$ by hypothesis. Thus by 2.1, $H(R, \mathbb{Z}_p) \cong \mathbb{Z}_p$, and there is essentially (i.e. up to multiplication with units of $\mathbb{Z}_p$) only one $\mathbb{Z}_p$-extension of K which is a field. (If S lies in $pH(R, \mathbb{Z}_p)$, the $\mathbb{Z}_p$-extension S cannot be connected since it is induced from a Galois extension of a proper subgroup of the Galois group.) This unique $\mathbb{Z}_p$-extension of K must then coincide with the cyclotomic $\mathbb{Z}_p$-extension Z/K. Moreover, W maps onto $\mathrm{Aut}(Z/K) \cong \mathbb{Z}_p$, and the kernel must be F since $W \cong \mathbb{Z}_p \oplus F$. Hence $\delta_K = p$-valuation of $|\mathrm{Ker}(W \longrightarrow \mathrm{Aut}(Z/K))|$.

By Serre (1978) [Lemma 2.9] or Coates (1977) [Appendix, Lemma 8], the stated formula follows.

Remark. It is conjectured that the expression in the theorem has the same p-adic valuation as the residue of the p-adic ζ-function of K at 1. This is known for $K/\mathbb{Q}$ abelian; cf. Coates (loc. cit.) or Leopoldt (1962).

Corollary 3.3. *If $K/\mathbb{Q}$ is totally real and unramified at p, then*

$$q_K \text{ has the same } p\text{-adic valuation as } h_K R_{K,p} \cdot p^{1-[K:\mathbb{Q}]}.$$

Proof. It suffices to treat the case $q_K \neq \infty$. Since $K/\mathbb{Q}$ is unramified in p, the p-th root of unity ζ_1 is not in K_v for all $v|p$, hence m_K as well as all m_v $(v|p)$ are zero. Hence $q_K = p^{\delta_K}$. Again since $K/\mathbb{Q}$ does not ramify in p, the number m_{K_1} is 1. Hence by the theorem:

$$(p\text{-adic val. of } q_K) - 1 = p\text{-adic val. of } h_K R_{K,p} \cdot \Delta^{-1/2} \cdot \prod_{v|p} (1 - \mathrm{N}v^{-1}).$$

Now Δ is a p-adic unit (again, because p does not ramify in K), and $\mathrm{N}v = p^{f_v}$, with f_v the residual degree of v over p. We have $\sum_{v|p} f_v = \sum_{v|p} [K_v : \mathbb{Q}_p] = [K:\mathbb{Q}]$, hence $\prod_{v|p} (1 - \mathrm{N}v^{-1})$ has the p-valuation $-[K:\mathbb{Q}]$. This concludes the proof.

As an easy example, we take $K = \mathbb{Q}$. Here the p-adic regulator $R_{K,p}$ is 1 by definition, therefore q_K is also 1 by the corollary. Hence $|F|_p = 1$, and we obtain $P(\mathbb{Z}[p^{-1}], C_{p^n}) = 1$ for all $n \geq 1$. In other words: All cyclic extensions of p-power degree of $\mathbb{Z}[p^{-1}]$ have a normal basis. Combining this with the discussion following III 3.6, we get: All cyclic extensions of p-power degree of $\mathbb{Q}$ which only ramify in p are cyclotomic. This is again a special case of the Kronecker–Weber theorem, but this time not too far away from the full theorem.

Now we turn to the case of CM fields. We briefly recall some properties of these: A number field K is called a CM field (CM standing for *complex multiplication*) if K is a quadratic totally imaginary extension of a totally real field (denoted K^+). Then there is a (unique) $j \in \mathrm{Aut}(K/\mathbb{Q})$ which induces complex conjugation in *every* embedding $K \to \mathbb{C}$. K^+ is the fixed field of j. For all abelian groups $A(K)$

which are functorially associated to number fields (such as unit groups and class groups) and on which multiplication by 2 is bijective, one has the canonical "plus and minus" decomposition:

$$A(K) = A(K)^+ \oplus A(K)^-, \quad \text{(also written } A^+(K) \oplus A^-(K))$$

with $A(K)^+ = \{a \in A \mid j(a) = a\} = (1+j)A(K)$, and $A(K)^- = \{a \in A \mid j(a) = -a\} = (1-j)A(K)$. In particular, j acts on $\mathrm{H}(R, C_{p^n})$ and $\mathrm{NB}(R, C_{p^n})$, and since $p \neq 2$, we obtain a decomposition

$$\mathrm{H}(R, C_{p^n}) = \mathrm{H}^+(R, C_{p^n}) \oplus \mathrm{H}^-(R, C_{p^n}),$$

and a corresponding decomposition of $\mathrm{NB}(R, C_{p^n})$.

Lemma 3.4. *The inclusion $K^+ \to K$ induces isomorphisms α, α':*

$$\mathrm{H}(R^+, C_{p^n}) \xrightarrow{\alpha} \mathrm{H}^+(R, C_{p^n}),$$

$$\mathrm{NB}(R^+, C_{p^n}) \xrightarrow{\alpha'} \mathrm{NB}^+(R, C_{p^n}). \qquad (\textit{Here } R^+ = \mathcal{O}_{K^+}[p^{-1}].)$$

Proof. We first prove the assertion for H. There is a map α: $\mathrm{H}(R^+, C_{p^n}) \longrightarrow \mathrm{H}(R, C_{p^n})$ defined by base extension. Exactly as in Lemma I 3.1 one sees that $\mathrm{Im}(\alpha)$ is fixed under j. (R/R^+ need not be a Galois extension, but this does not matter.) There is a corresponding map on the field level α_K: $\mathrm{H}(K^+, C_{p^n}) \longrightarrow \mathrm{H}^+(K, C_{p^n})$. Now α_K is (in cohomological terms) just restriction along $\Psi_K \subset \Psi_{K^+}$, and one has the corestriction cor: $\mathrm{H}^1(\Psi_K, C_{p^n}) \to \mathrm{H}^1(\Psi_{K^+}, C_{p^n})$ going the other way. Also, $\mathrm{H}(R^+, C_{p^n})$ is identified with a subgroup of $\mathrm{H}(K^+, C_{p^n})$ by **0** 4.2, and similarly $\mathrm{H}^+(R, C_{p^n}) \subset \mathrm{H}^+(K, C_{p^n})$.

<u>α is injective:</u> It suffices to show α_K injective. But $\mathrm{cor}\cdot\alpha_K$ is multiplication by 2, hence bijective, so α_K is injective.

<u>α is surjective:</u> Let $B \in \mathrm{H}(R, C_{p^n})$ with $j(B) = B$. Consider $A = \mathrm{cor}(B) \in \mathrm{H}(K^+, C_{p^n})$. Then $\alpha_K(A) = \alpha_K\mathrm{cor}(B) = (1+j)(B)$ (since corestriction followed by restriction is multiplication by the norm element in $\mathbb{Z}[\mathrm{Aut}(K/K^+)]$), and the last term is $2\cdot B$ since $j(B) = B$. Hence $B \in \mathrm{Im}(\alpha_K)$. Now it is not hard to check using ramification indices and $p \neq 2$ that *any* preimage A of B must also be unramified outside primes over p (since B is), so B is actually in the image of α.

It remains to prove the assertion concerning NB. This is proved the same way, but there is a complication: we need to know that if $\alpha(A)$ has normal basis over R, then A has normal basis over R^+. We consider the base extension map β: $\mathrm{Pic}(R^+[C_{p^n}]) \longrightarrow \mathrm{Pic}(R[C_{p^n}])$. Abbreviate: $T^+ = R^+[C_{p^n}]$, $T = R[C_{p^n}]$. Suppose I is an invertible ideal of T^+ such that $T\cdot I$ is free, say with generator x. Then (since $j(TI) = TI$) there is $u \in T^*$ with $j(x) = u\cdot x$.. Then $u\cdot j(u) = 1$, and with $v = j(u)$ one gets $j(v)/v = u/j(u) = u^2$. Hence $j(x^2/u) = x^2/u$, i.e. the ideal TI^2 has a j-invariant generator $x^2/v \in T^+$. By descent, I^2 is the principal ideal generated by x^2/v. Hence

$2\mathrm{Ker}(\beta) = 0$. Since the Picard invariant pic maps $\mathrm{H}(R^+, C_{p^n})$ into the p^n-torsion of $\mathrm{Pic}(R^+[C_{p^n}])$, we obtain that $\mathrm{pic}(A)$ is trivial, q.e.d.

From this lemma one sees that $|\mathrm{P}(R, C_{p^n})^+|$ converges with $n \to \infty$ to q_{K^+}. Therefore, if q_K and q_{K^+} are both assumed to be finite, then $|\mathrm{P}(R, C_{p^n})^-| = |\mathrm{P}(R, C_{p^n})| \cdot |\mathrm{P}(R, C_{p^n})^+|^{-1}$ converges to q_K/q_{K^+}, which we also shall write q_K^-. From now on, we shall deal exclusively with q_K^-. (Note that we have done something very common in algebraic number theory: we decomposed a "class number" into two factors, the first of which comes from the maximal real subfield.)

For the next theorem, recall that $h_K^- = h_K/h_{K^+}$, and h_K^- is indeed the order of the group $\mathrm{Cl}(K)^-$.

Theorem 3.5. *For every CM-field K, the orders $|\mathrm{P}(R, C_{p^n})^-|$ converge unconditionally to a value q^* (which is q_K^- if both q_K and q_{K^+} are finite), and q^* is a divisor of h_K^-.*

Proof. We use the standard method of comparing the situation for K and K^+. We first calculate $|\mathrm{NB}^-(R, C_{p^n})|$ as the quotient of $a_n = |\mathrm{NB}(R, C_{p^n})|$ by $a_n^+ = |\mathrm{NB}^+(R, C_{p^n})|$. To this end, let $\Sigma = \{$places of K dividing $p\}$ and $\Sigma^+ = \{$places of K^+ dividing $p\}$. If $w \in \Sigma$, w divides $v \in \Sigma^+$, we have the following possibilities:

1) v splits in K/K^+, or

2) v is inert in K/K^+, or

3) v ramifies in K/K^+.

Over any v of type 1) there are two $w \in \Sigma$; over any v of type 2) or 3), there is just one w. If v is of type 1), and $w|v$, then $K^+_v = K_w$, hence $m_v = m_w$. In cases 2) and 3) one sees by degree considerations: either $m_v = m_w$ or $[m_v = 0$ and $m_w = 1]$. The latter case will be called the exceptional case; let $\varkappa$ be the number of $v \in \Sigma^+$ for which it occurs. Observe that m_{K^+} is certainly 0 since K^+ is real. For $n \geq \max(m_w | w \in \Sigma)$ we obtain by III 3.6 using that $s(K^+) = 0$:

$$a_n = p^{-m_K + \sum_{w\in\Sigma} m_w} \cdot p^{n(s+1)}, \quad \text{and}$$
$$a_n^+ = p^{\sum_{v\in\Sigma^+} m_v} \cdot p^n .$$

Letting $\Sigma' \subset \Sigma^+$ denote the set of split v, we obtain by dividing the first equation by the second:

$$a_n/a_n^+ = p^{-m_K + \sum_{v\in\Sigma'} m_v + \varkappa} \cdot p^{ns}.$$

In the next step, we calculate $|\mathrm{H}^-(R, C_{p^n})|$, again as a quotient. Recall $\mathrm{H}(R, C_{p^n}) \cong \mathrm{Hom}(W, C_{p^n})$ where W was defined in §1. Then just as well $\mathrm{H}^-(R, C_{p^n}) \cong \mathrm{Hom}(W_{(p)}^-, C_{p^n})$ with $W_{(p)}$ the pro-p-part of W. (Then 2 operates bijectively on $W_{(p)}$, and it is legal to form $W_{(p)}^-$.) We now recall the exact sequence of §1:

$$1 \longrightarrow \Big(\prod_{w\in\Sigma} U_w\Big)\Big/\overline{\mathrm{Im}(E(K))} \longrightarrow W \longrightarrow \mathrm{Cl}(K) \longrightarrow 1.$$

Abbreviate the term between 1 and W by V. Take pro-p-parts and then minus parts in this short exact sequence of profinite groups. One obtains another sequence

$$1 \longrightarrow V_{(p)}^{-} \longrightarrow W_{(p)}^{-} \longrightarrow \mathrm{Cl}(K)_{(p)}^{-} \longrightarrow 1.$$

Now the minus part of $\overline{\mathrm{Im}(E(K))}_{(p)}$ is precisely $\mathrm{Im}(\mu_K)_{(p)}$, since a) j maps every root ζ of unity to ζ^{-1}, so $\mathrm{Im}(\mu_K)_{(p)}$ is in the minus part, and b) the index of $E(K^+)\cdot\mu_K$ in $E(K)$ is 1 or 2 by Washington (1982) [Thm. 4.12] and $p \neq 2$. (The cited result is not a difficult one.) Hence $V_{(p)}^{-}$ is the pro-p-part of $\big(\prod_{w\in\Sigma} U_w\big)/\mathrm{Im}(\mu_K)$.

We want to calculate the torsion part of $V_{(p)}^{-}$. To this end, we write

$$\Big(\prod_{w\in\Sigma} U_w\Big)^{-} = \prod_{v\in\Sigma^+}\Big(\prod_{w|v} U_w\Big)^{-},$$

and we separate the cases:

(1) If v splits in K, then there are exactly two places dividing v in K, say w and w', and $K_w = K_{w'} = K^+_v$. One sees then that $(U_w\times U_{w'})^{-} \cong U_v$, and therefore $\mathrm{tors}\,(U_w\times U_{w'})^{-} \cong \mu(K_v)$.

(2) and (3): v inert or ramified in K. Then there is just one w over v; $\mathrm{tors}(K_w)^{-} \cong \mathrm{tors}(K_w)/\mathrm{tors}(K_w)^{+}$ (except maybe in the 2-component), and $\mathrm{tors}(K_w)^{+} = \mathrm{tors}(K_v)$. Hence the p-primary torsion in U_w^{-} is trivial if $m_w = m_v$, and of order p if we are in the exceptional case.

Putting this together, we have: The p-primary torsion of $\big(\prod_{w\in\Sigma} U_w\big)^{-}$ is of order p^e, $e = \sum_{v\in\Sigma'} m_v + \varkappa$. If we now factor out $\mathrm{Im}(\mu_K)$, we obtain that the p-primary torsion of V^{-} is of order p^{e-m_K}. It is an elementary consequence of this and the above sequence that the torsion in the pro-p-part of W^{-} has order $p^{e-m_K}\cdot q^*$, with q^* a divisor of h_K^{-}. It is also easily checked that the $\mathbb{Z}_p$-rang of V^{-} (and of W^{-}) equals $s = s(K)$. Using that $\mathrm{H}^{-}(R, C_{p^n}) \cong \mathrm{Hom}(W_{(p)}^{-}, C_{p^n})$, we find for all n:

$$|\mathrm{H}^{-}(R, C_{p^n})| \;\text{ divides }\; p^{ns}\cdot p^{e-m_K}\cdot q^*,$$

with equality for $p^n \ge p^{e-m_K}\cdot q^*$. We now divide this by the formula $|\mathrm{NB}^{-}(R, C_{p^n})| = p^{ns}\, p^{e-m_K}$ found above, and end up with

$$|\mathrm{NB}^{-}(R, C_{p^n})| \;\text{ divides }\; q^* \qquad \text{for } n \ge \max_{w\in\Sigma}(m_w),$$

with equality for $p^n \ge p^{e-m_K}\cdot q^*$. This concludes the proof: q^* divides h_K^{-}, and $\lim_{n\to\infty} |\mathrm{NB}^{-}(R, C_{p^n})| = q^*$.

It is natural to ask how the number q^* is related to the index q_∞^{-} of $\mathrm{NB}^{-}(R, \mathbb{Z}_p)$ in $\mathrm{H}^{-}(R, \mathbb{Z}_p)$. There is the exact sequence

$$0 \longrightarrow \mathrm{NB}^-(R, C_{p^n}) \longrightarrow \mathrm{H}^-(R, C_{p^n}) \longrightarrow \mathrm{P}^-(R, C_{p^n}) \longrightarrow 0,$$

and this sequence stays exact on applying $\lim_{\leftarrow}$ since all groups $\mathrm{NB}^-(R, C_{p^n})$ are finite and hence satisfy the Mittag-Leffler property (Jensen (1972)). Hence q_∞^- is just the order of $\lim_{\leftarrow} \mathrm{P}^-(R, C_{p^n})$. As an easy consequence we obtain:

Theorem 3.6. *The index* $q_\infty^- = [\mathrm{H}^-(R, \mathbb{Z}_p) : \mathrm{NB}^-(R, \mathbb{Z}_p)]$ *divides* q^* *(in particular, it is finite). If the maps* $\mathrm{P}^-(R, C_{p^{n+1}}) \longrightarrow \mathrm{P}^-(R, C_{p^n})$ *are isomorphisms for n large, then* $q_\infty^- = q^*$.

We do not know whether the last mentioned condition is always fulfilled. In the next section we shall discuss examples K of small degree over $\mathbb{Q}$.

§ 4 Calculation of q_K: examples

Let first K be an imaginary quadratic extension of $\mathbb{Q}$. Then K is a CM-field with $K^+ = \mathbb{Q}$. Since $q_{\mathbb{Q}} = 1$ as shown in §3 (or, as follows from the Kronecker-Weber theorem), we have $q_K = q_K^-$ (see §3). We also know by 3.5 that q_K^- divides h_K. Observe that $h_K = h_K^-$ since $h_{\mathbb{Q}} = 1$.

It is elementary to see that the maximum of m_w, w place of K over p, is at most 1 (and 0 for $p > 3$). Hence we get by the end of the proof of 3.5:

$$|\mathrm{P}(R, C_p)| \quad \text{divides } q_K.$$

To find fields with $q_K > 1$, it therefore suffices to find examples K with $\mathrm{P}(R, C_p)$ nontrivial. By I 2.7 and I 3.6, $\mathrm{P}(R, C_p) \cong \mathrm{Pic}(S_1)[p](-1)^{\Gamma_1}$. For $p = 3$, this can be simplified using a sort of Spiegelungsprinzip, as in the proof of Scholz' theorem:

Suppose $p = 3$ and (for simplicity) $K \neq \mathbb{Q}(\sqrt{-3})$. Then $\Gamma_1 = \{1, \sigma\}$ with $\sigma(\zeta) = \zeta^{-1}$ (ζ a primitive 3rd root of unity), $\sigma|K = id_K$. We also need the automorphism j of $K_1 = K(\zeta)$ which acts as complex conjugation on K and as identity on ζ. From the above we obtain:

$$\mathrm{P}(R, C_p) = \mathrm{P}^-(R, C_p) = \mathrm{P}(R, C_p)^{1+j} \quad \text{(the exponent means: subgroup annihilated by ..)}$$

$$\cong \left(\mathrm{Pic}(S_1)[p](-1)^{\Gamma_1}\right)^{1+j}.$$

Now the fixed group under Γ_1 is just the kernel of $1-\sigma$. If we take the twist into account ($\omega(\sigma) = -1$), we can rewrite the last expression as

$$\left(\mathrm{Pic}(S_1)[p]^{1+\sigma}\right)^{1+j} =: A.$$

Consider the biquadratic extension $K(\zeta)/\mathbb{Q}$. Its Galois group is a "Klein four-group" with elements $1, \sigma, j, j\sigma$. We can rewrite A as follows:

$$A = \left(\mathrm{Pic}(S_1)[p]^{1-j\sigma}\right)^{1+j}.$$

But now $j\sigma$ is the complex conjugation of the CM-field $K(\zeta)$, i.e. $\mathrm{Pic}(S_1)[p]^{1-j\sigma} = \mathrm{Pic}(S_1)^+[p]$. It is well-known that $\mathrm{Pic}(S_1)^+[p] \cong \mathrm{Pic}(S_1^+)[p]$ (S_1^+ being the ring of p-integers in $K(\zeta)^+$) since $3 = p \neq 2$. Finally, j induces the nontrivial automorphism of $K(\zeta)^+$, hence $1+j$ induces the norm from $\mathrm{Pic}(S_1^+)$ to $\mathrm{Pic}(\mathbb{Z}[p^{-1}])$, and the latter Picard group is zero. We end up with

$$P(R, C_p) \cong A \cong \mathrm{Pic}(S_1^+)[p] \qquad (\text{recall } p = 3).$$

We have thus proved:

Theorem 4.1. *If $K/\mathbb{Q}$ is imaginary quadratic, $p = 3$, and the Picard group $\mathrm{Pic}(S_1^+)$ of the ring of p-integers in $K(\zeta)$ ($\zeta^3 = 1, \zeta \neq 1$) has nontrivial 3-torsion, then $q_{K,p} > 1$.*

Examples. If $K = \mathbb{Q}(\sqrt{-d})$ ($d \in \mathbb{N}$ square-free), then $K(\zeta)^+ = \mathbb{Q}(\sqrt{3d})$. If 3 is inert or ramified in $K(\zeta)^+$, then the 3-part of $\mathrm{Pic}(S_1^+)$ is isomorphic to the 3-part of $\mathrm{Pic}(O_{K(\zeta)^+})$. In particular, to guarantee the hypotheses of 4.1 it is sufficient to have: 3 does not divide d (then 3 ramifies in $K(\zeta)^+ = \mathbb{Q}(\sqrt{3d})$), and the class number of $\mathbb{Q}(\sqrt{3d})$ is divisible by 3. Consulting a class number table, one finds that the smallest such d is $d = 107$. One can also replace these two conditions by [$d = 3d'$ with $d' \equiv -1 \pmod 3$, and the class number of $\mathbb{Q}(\sqrt{d'})$ is divisible by 3], since then 3 is inert in $K(\zeta)^+ = \mathbb{Q}(\sqrt{d'})$. The first six values of such d' are $d' = 254, 257, 326, 359, 443, 473$. The case $d' = 257$, i.e. $K = \mathbb{Q}(\sqrt{-3\cdot 257})$, has also been treated by Kersten and Michaliček, using results of Greenberg. We shall get back to this.

Remark. Since $q_K = q_K^-$ divides $h_K = h_K^-$ by Thm. 3.5, we have reproved a part of Scholz' theorem, see Washington (1982) [Thm. 10.10]. To wit, we have proved: if 3 divides $h(\mathbb{Q}(\sqrt{3d}))$ and 3 is inert or ramified in $\mathbb{Q}(\sqrt{3d})$, then 3 also divides $h(\mathbb{Q}(\sqrt{-d}))$.

Now let K be any CM-field, $p > 2$ again arbitrary. For simplicity we assume that $[K(\zeta):K] = p-1$. (ζ is a p-th primitive root of unity.) We are interested in the index q^-_∞ of $\mathrm{NB}(R, \mathbb{Z}_p)$ in $\mathrm{H}(R, \mathbb{Z}_p)$. We repeat (3.6): q^-_∞ divides $q^- = q_K^-$, but it is not clear whether they are equal.

Let $E = K(\zeta)^+$ and $E_n = K(\zeta_n)^+$ with ζ_n a primitive p^n-th root of 1. These fields are totally real, and $\bigcup E_n = E_\infty$ is a $\mathbb{Z}_p$-extension of E. We suppose here that the reader is somewhat familiar with Iwasawa theory: There exist $\lambda, \mu \in \mathbb{N}$ and $\nu \in \mathbb{Z}$ such that

$$\text{order of the } p\text{-primary part of } \operatorname{Pic}(\mathcal{O}_{E_n}[p^{-1}] = p^{(p^{\mu n}+\lambda n+\nu)}$$

for n large. (Iwasawa (1973)). Obviously then the left hand side is bounded iff μ and λ are both zero; if this is the case, the left hand side is eventually constant $(= p^{\nu})$.

Theorem 4.2. *With K and E_n as above, we have: If the order of the p-primary part of $\operatorname{Pic}(\mathcal{O}_{E_n}[p^{-1}]$ is bounded for $n \to \infty$, then $q^-_\infty = q^-$.*

Proof. As seen in Thm. 3.6, we must show that the map $P(R, C_{p^{n+1}}) \longrightarrow P(R, C_{p^n})$ is an isomorphism for large n. We have $P(R, C_{p^n}) \cong \operatorname{Pic}(S_n)[p^n](-1)^{\Gamma_n}$ with $S_n = \mathcal{O}_{E_n}[p^{-1}]$ the ring of p-integers in $K_n = K(\zeta_n)$ as usual. Exactly as in the proof of 4.1, one shows (using the element $\sigma \in \Gamma_n$ with $\omega(\sigma) = -1$, and $j \in \operatorname{Aut}(K_n/\mathbb{Q})$ with $j|K =$ conjugation, $j(\zeta_n) = \zeta_n$): The minus part of $P(R, C_{p^n})$ is canonically isomorphic to

$$A_n = \operatorname{Pic}(S_n^+)[p^n](-1)^{\Gamma_n} \qquad (S_n^+ = p\text{-integers in } E_n).$$

The maps $P(R, C_{p^{n+1}}) \longrightarrow P(R, C_{p^n})$ correspond to the maps $N_n\colon A_{n+1} \to A_n$ induced by the norm from E_{n+1} to E_n, by the remark following 2.4. If these norm maps $N_n\colon \operatorname{Pic}(S_{n+1}^+) \longrightarrow \operatorname{Pic}(S_n^+)$ are isomorphisms for $n >> 0$, then $|\operatorname{Pic}(S_n^+)|$ is bounded and $\operatorname{Pic}(S_n^+) = \operatorname{Pic}(S_n^+)[p^n]$ for n large. By the same token, one then obtains that $N_n\colon A_{n+1} \to A_n$ is an isomorphism for large n, since N_n is compatible with the Γ_{n+1}- and Γ_n-actions in an obvious sense. Hence it suffices to show: $N_n\colon \operatorname{Pic}(S_{n+1}^+) \longrightarrow \operatorname{Pic}(S_n^+)$ is an isomorphism for n large.

Now we see from the hypothesis that it is enough to show N_n surjective for n large (since $|\operatorname{Pic}(S_n^+)|$ is eventually constant). This in turn is a well-known consequence of the fact that in a $\mathbb{Z}_p$-extension, from some step on, all divisors of p are totally ramified (see Washington (1982) [Lemma 13.3 and Thm. 10.1]). q.e.d.

The motivation for this theorem is as follows. It is widely believed (although unproved) that the hypothesis "$\lambda = \mu = 0$" of the theorem is satisfied for the cyclotomic $\mathbb{Z}_p$-extension E_∞ of an *arbitrary* totally real number field E.

By a calculation of Greenberg (1977), a nontrivial example is available: Let $p = 3$, $K = \mathbb{Q}(\sqrt{-3\cdot 257})$ (we've encountered this field already) and $E_n = K_n^+ = \mathbb{Q}(\sqrt{257}, \zeta_n)^+$, and S_n^+ the ring of p-integers in E_n. The class number of all E_n is three, and hence also the order of $\operatorname{Pic}(S_n^+)$ is three (one can show that the ideal above 3 in E_n is principal). Hence, by the theorem, we have $q_K = q_{K,\infty}$. We know that 3 divides q_K (see above); on the other hand, the order of each $P(R, C_{p^n})$ is majorized by $|\operatorname{Pic}(S_n^+)|$ by the proof of 4.3. Hence:

Corollary 4.3. *For $K = \mathbb{Q}(\sqrt{-3\cdot 257})$ and $p = 3$, we have $q_K = q_{K,\infty} = 3$.*

(Cf. Kersten and Michaliček (1989b), Example 2.6.)

To conclude this section we briefly look at the totally real case. We first remark that it is easy to calculate the p-adic regulator for real quadratic fields K (it suffices to know the fundamental unit). Hence one may calculate q_K using Cor. 3.3. It is left to the interested reader to compute numerical examples.

Let now K be any totally real number field. It seems likely that q_K can be "anything". But something can be said about $q_{K,\infty}$:

Proposition 4.4. *For K totally real, $q_{K,\infty} = \infty$ or $q_{K,\infty} = 1$.*

Proof. Suppose $q_{K,\infty} \neq \infty$, i.e. $\mathrm{NB}(R,\mathbb{Z}_p)$ and $\mathrm{H}(R,\mathbb{Z}_p)$ have the same $\mathbb{Z}_p$-rank (= $s+1$ by Thm. 2.2). Now $s = 0$, so $\mathrm{H}(R,\mathbb{Z}_p) \cong \mathbb{Z}_p$, whence $\mathrm{NB}(R,\mathbb{Z}_p) = p^a \cdot \mathrm{H}(R,\mathbb{Z}_p)$ for some $a \geq 0$. If $a > 0$, then any $B \in \mathrm{NB}(R,\mathbb{Z}_p)$ is a multiple of p, i.e. of the form $\iota^* C$, where ι: $\mathbb{Z}_p \to \mathbb{Z}_p$ is multiplication by p. As already used earlier, this forces B (to be precise, already the first layer $B_1 \in \mathrm{NB}(R,C_p)$) to have nontrivial idempotents. Contradiction, since we may take the cyclotomic $\mathbb{Z}_p$-extension $Z = \bigcup A_n(R)$ for B, and Z is connected, with normal basis. (I §1) Hence $a = 0$, which means $q_{K,\infty} = 1$.

The same argument proves:

Corollary 4.5. *For K totally real, $NB(R,\mathbb{Z}_p)$ is spanned by the cyclotomic $\mathbb{Z}_p$-extension Z as a $\mathbb{Z}_p$-module.*

§5 Torsion points on abelian varieties with complex multiplication

We have seen that cyclotomic p-power extensions of a number field K give extensions A_n/R with normal basis ($R = \mathcal{O}_K[p^{-1}]$ as always). Of course, these extensions are obtained by adjoining values of the exponential function at torsion points of the additive group $\mathbb{C}/2\pi i\mathbb{Z}$. It seems natural to look at the other vast class of abelian extensions that can be obtained, very roughly speaking, by adjoining torsion points of appropriate groups, namely extensions defined by the adjunction of torsion points of abelian varieties of CM type.

In this section we suppose the reader has some acquaintance with the latter theory. Proofs will be sketchier than in other partes of these notes. We have to admit that it would be more satisfying to *construct explicitly* extensions with normal basis using torsion points in abelian varieties (as done by Cassou-Noguès and Taylor (1986) for elliptic curves). This we cannot do; we use all our knowledge about extensions with normal basis, and considerable knowledge on the extensions gotten from torsion points on abelian varieties, to deduce a connection between these two theories. We use the notation of Lang (1983) unless otherwise stated.

Let A be an abelian variety defined over the number field k with complex multiplication by the ring of integers $\mathcal{O} = \mathcal{O}_{K'}$ of the CM field K', i.e. we have an injection of rings $\mathcal{O} \longrightarrow \mathrm{End}_k(A)$ and $2\cdot\dim(A) = [K':\mathbb{Q}]$. (It is always helpful to think of the example A an elliptic curve, and $K'/\mathbb{Q}$ imaginary quadratic.) Let Φ' be the CM type of K' which is given by the embedding $\mathcal{O} \to \mathrm{End}_k(A)$, and let K be the *reflex field* of Φ'. (We will also say that K is the reflex of K'.) Then K lies in k (see Lang; we have interchanged the roles of K and K', which simplifies notation for us). K is again a CM-field; let $2s$ denote its degree over $\mathbb{Q}$. (If K' is imaginary quadratic over $\mathbb{Q}$, then $K = K'$.) Finally, let j be the (!) complex conjugation on K. Assume $p \neq 2$.

Definition. We call a $\mathbb{Z}_p$-extension L_∞/K

negative, if $j[L_\infty] = -[L_\infty]$ in the $\mathbb{Z}_p[\{1,j\}]$-module $\mathrm{H}(K,\mathbb{Z}_p)$;

of A-type, if $kL_\infty \subset k(A_{tors})$.

The second notion, as it stands, makes only sense if L_∞ is a field. If L_∞ is not a field, but also not the trivial $\mathbb{Z}_p$-extension, then $L_\infty = [p^n]^* L'_\infty$ for a field extension L'_∞, and we define: L_∞ is of A-type if L'_∞ is.

Now let $R = \mathcal{O}_K[p^{-1}]$ as always, B_∞ = ring of p-integers of L_∞. (Hence $B_\infty \in \mathrm{H}(R,\mathbb{Z}_p)$ since $\mathbb{Z}_p$-extensions are unramified outside p.) We would like to know whether a nontrivial $\mathbb{Z}_p$-multiple of B_∞ is in $\mathrm{NB}(R,\mathbb{Z}_p)$. By III §3 this is certainly the case if the Leopoldt conjecture holds for K and p. Therefore, let us not assume the validity of Leopoldt's conjecture. The result will then be, quite roughly: Up to nonzero factors in $\mathbb{Z}_p$, the $\mathbb{Z}_p$-extensions of R with normal bases are the same as those that can be obtained by adjoining torsion points on suitable abelian varieties with CM. Invoking again the analogy between torsion points on such varieties, and roots of unity, we might say that the $\mathbb{Z}_p$-extensions coming from CM theory are an analog of the cyclotomic $\mathbb{Z}_p$-extension in the real case (cf. Cor. 4.5). More precisely, we have:

Theorem 5.1. *Let K be a number field and L_∞/K be a $\mathbb{Z}_p$-extension, L_∞ a field. Consider the following three conditions:*

a) L_∞ is of A-type for A some abelian variety defined over some $k \subset K$, with complex multiplication by $\mathcal{O}_{K'}$, such that K is the reflex of K'. (NB. This includes the condition $2\dim(A) = [K':\mathbb{Q}]$.)

b) In the group $\mathrm{H}(K,\mathbb{Z}_p)$, L_∞ is the product of a negative extension and a $\mathbb{Z}_p$-multiple of the cyclotomic $\mathbb{Z}_p$-extension.

c) Some nonzero $\mathbb{Z}_p$-multiple of B_∞ (= p-integers of L_∞) is in $\mathrm{NB}(R,\mathbb{Z}_p)$.

Then b) $\Longleftrightarrow$ c); a) $\Rightarrow$ b); and if K possesses at least one nondegenerate CM *type, then also b) $\Rightarrow$ a).*

Remark. We shall not explain nondegenerate types here. For their definition, see Lang(1983) [ch. 6]. For existence questions, consult loc. cit., Schappacher (1977), and Schmidt (1984).

Proof of **5.1.** b) $\Longleftrightarrow$ c): This is fairly easy. Let $W = W_K$ be the Galois group of the maximal p-abelian extension of K unramified outside p. (We have seen this pro-p-group before. It is isomorphic to $I_K/(K^*\cdot\prod_{\mathfrak{v}\nmid p} U_{\mathfrak{v}})$.) Let L_∞ correspond to $f \in \mathrm{Hom}(W, \mathbb{Z}_p)$. Since $p \neq 2$, we may write $f = f^+ + f^-$ with $f^\pm \in \mathrm{Hom}(W^\pm, \mathbb{Z}_p)$. As shown in Thm. 3.6, the index of $\mathrm{NB}(R,\mathbb{Z}_p)^-$ in $\mathrm{H}(R,\mathbb{Z}_p)^-$ is finite and divides q_K^-, hence $q_K^-\cdot f^-$ belongs to an extension with normal basis, and it remains to show: A plus extension, i.e. an element of $\mathrm{H}(R,\mathbb{Z}_p)^+$, has normal basis iff it is a multiple of the cyclotomic $\mathbb{Z}_p$-extension. This has been shown before, see Cor. 4.5.

a) $\Rightarrow$ b): Let L_∞/K correspond to $f\colon W_K \longrightarrow \mathbb{Z}_p$ (W_K as in the last paragraph), and assume $kL_\infty \subset k(A_{tors})$. Let Φ be the CM type of K, i.e. the dual type to Φ' above, and N_Φ be the type norm $K \to K'$. We need a lemma.

Lemma 5.2. *With the above notations, and for any $\mathbb{Z}_p$-extension M_∞/k with $M_\infty \subset k(A_{tors})$, given by $g\colon W_k \to \mathbb{Z}_p$, the map g annihilates the group*

$$\mathrm{Ker}(\mathrm{N}_\Phi \circ \mathrm{N}_{k/K}) \quad (\textit{with norm maps } W_k \xrightarrow{\mathrm{N}_{k/K}} W_K \xrightarrow{\mathrm{N}_\Phi} W_{K'}).$$

(NB. K is indeed a subfield of k, not the other way!)

Proof. By the second main theorem of complex multiplication [Lang (1983) p.84], any idele $s \in I_k$ acts (via the Artin map) on A_{tors} exactly as $\mathrm{N}_\Phi \mathrm{N}_{k/K}(s^{-1})\cdot\alpha(s)$ acts on A_{tors} via the given "complex multiplication", i.e. via the given embedding of $\mathcal{O}_{K'}$ into $\mathrm{End}(A)$. Here $\alpha\colon I_k \longrightarrow K'^*$ is the so-called CM *character associated to A*. (Note: The element $y = \mathrm{N}_\Phi \mathrm{N}_{k/K}(s^{-1})\cdot\alpha(s)$ is a unit idele of K', not an integer of K'. But we may regard y as a unit of $\hat{\mathbb{Z}}\otimes_{\mathbb{Z}}\mathcal{O}_{K'}$, so y does act on A_{tors}.)

We know moreover that the group $\prod_{\mathfrak{v}|p} U(k_{\mathfrak{v}}) \subset I_k$ has a subgroup V of finite index on which α is trivial. Then V projects to a subgroup V' of finite index in W_k. For any $s' \in V'$ (with preimage $s \in V$), we have $\mathrm{N}_\Phi \mathrm{N}_{k/K}(s'^{-1})$ = image of $\mathrm{N}_\Phi \mathrm{N}_{k/K}(s^{-1})$ in W_k = image of $\mathrm{N}_\Phi \mathrm{N}_{k/K}(s^{-1})\cdot\alpha(s)$ in W_k. Therefore, if $s' \in W_k$ is in the kernel of $\mathrm{N}_\Phi \mathrm{N}_{k/K}$, and in V', then s' acts trivially (via the Artin map) on A_{tors}, and in particular on M_∞. With $m = [W_k{:}V']$, we therefore have that mg kills the kernel of $\mathrm{N}_\Phi \mathrm{N}_{k/K}$. Since the range $\mathbb{Z}_p$ of g is torsion-free, the conclusion of the lemma follows.

Back to the proof of a) $\Rightarrow$ b): If L_∞/K is given by $f\colon W_K \to \mathbb{Z}_p$, then the extension $M_\infty = k\otimes_K L_\infty$ is given by $g = f\circ \mathrm{N}_{k/K}\colon W_k \to \mathbb{Z}_p$, as follows from a standard property of the Artin symbol. We apply 5.2 to g and find that:

$$f \circ N_{k/K} \text{ annihilates } \mathrm{Ker}(N_\Phi N_{k/K}),$$

hence

$$f|_{\mathrm{Im}(N_{k/K})} \text{ annihilates } \mathrm{Ker}(N_\Phi).$$

By the standard argument "$N_{k/K} \circ$ inclusion = multiplication by $[k:K]$", the cokernel of $N_{k/K}$: $W_k \to W_K$ is torsion. Again, the range of f is torsion-free, so f kills the kernel of N_Φ. Now decompose $f = f^+ + f^-$ as above, with f^+: $W_K^+ \to \mathbb{Z}_p$. But on W_K^+ the type norm N_Φ is just one half of the restriction of the absolute norm $N_{K/\mathbb{Q}}$. This means: f^+ kills $\mathrm{Ker}(N_{K/\mathbb{Q}})$. Again $\mathrm{Coker}(N_{K/\mathbb{Q}}$: $W_K \to W_\mathbb{Q})$ is a bounded torsion group, hence a nontrivial multiple of f^+ factors through $N_{K/\mathbb{Q}}$, which means: A nonzero multiple mf^+ of f^+ comes from a multiple of the cyclotomic $\mathbb{Z}_p$-extension of $\mathbb{Q}$ (since *all* $\mathbb{Z}_p$-extensions of $\mathbb{Q}$ are of this kind), i.e. mf^+ itself corresponds to a multiple of the cyclotomic $\mathbb{Z}_p$-extension of K. From this it follows that f^+ itself is cyclotomic. This proves b).

b) $\Rightarrow$ a): Assume the stated hypotheses. We start with another lemma:

Lemma 5.3. *Let M_∞/k be a $\mathbb{Z}_p$-extension, $M_\infty \subset k(A_{tors})$, and $K \subset k$. Then some nontrivial multiple of M_∞ descends to K, i.e. there is a $\mathbb{Z}_p$-extension L_∞/K with $k \otimes_K L_\infty = m \cdot L_\infty$ for some $m \neq 0$.*

Proof. If f: $W_K \to \mathbb{Z}_p$ defines L_∞, then as above, $f N_{k/K}$ defines $k \otimes_K L_\infty$. Let now M_∞ be defined by g: $W_k \to \mathbb{Z}_p$. It suffices to show: For some $m \neq 0$ we can write $mg = f\, N_{k/K}$ for some f. By Lemma 5.2, g kills $\mathrm{Ker}(N_{k/K})$ since trivially this kernel is contained in $\mathrm{Ker}(N_\Phi N_{k/K})$. It suffices to take for m the exponent of the cokernel of $N_{k/K}$: $W_k \to W_K$ (this exponent can be estimated by $[k:K]$, as we have seen). Q.E.D.

Back to the proof of b) $\Rightarrow$ a): take some nondegenerate CM type Φ on K. Then one knows: There *exists* some abelian variety A, defined over some $k \supset K$, with complex multiplication by $\mathcal{O}_K$, and associated CM type Φ', where (K', Φ') is the reflex of (K, Φ). From the *nondegeneracy* of Φ and Φ' and the second theorem of C.M., one now infers that $k(A_{tors})$ contains at least $s+1$ independent $\mathbb{Z}_p$-extensions, where $2s = [K:\mathbb{Q}] = [K':\mathbb{Q}]$. (See Lang (1983) [IV 2.8] or Ribet (1980). We omit the details.) By 5.3 one then gets that K has at least $s+1$ independent $\mathbb{Z}_p$-extensions of A-type. Now one deduces easily from class field theory that there are exactly s independent negative $\mathbb{Z}_p$-extensions of K ("the minus part of Leopoldt's conjecture is true"), hence there are exactly $s+1$ independent $\mathbb{Z}_p$-extensions of K which satisfy b). Since a) $\Rightarrow$ b) is already proved, the converse must also hold. q.e.d.

§6 Further results: a short survey

We here assemble, without proofs or with scanty indications of proof, some other results in Galois theory of rings which have a relation to algebraic number theory. Some of these appear in the literature, a few others are unpublished.

It is appropriate to begin with the theory of Kersten and Michaliček. They were the first to obtain a description of $\mathrm{NB}(R, C_{p^n})$ for an arbitrary connected ring R which contains p^{-1} ($p \neq 2$, as usual in this theory). Their approach is in principle not totally different from the descent theory presented in these notes, but the methods and techniques differ considerably from ours. The most important technical tool they use is the group ring $R[C_{p^n}^{\wedge}]$ ($^{\wedge}$ means dual; the dual of a finite abelian group A is isomorphic to A, but not canonically). The main result in Kersten and Michaliček (1988) is an isomorphism of $\mathrm{NB}(R, C_{p^n})$ with a certain subquotient of the unit group of $R[C_{p^n}^{\wedge}]$. I feel that, by dint of probably quite long calculations, one can deduce our description of NB from this one, and vice versa, but I have not checked all of it. As a consequence, Kersten and Michaliček also proved Thm IV 2.2 of these notes for the case that K is a CM field. In a similar vein, they obtained lifting theorems comparable to ours in II §4. The quotient $\mathrm{P}(R, C_{p^n}) = (\mathrm{H/NB})(R, C_{p^n})$ seems to be less accessible by these methods.

Besides Leopoldt's conjecture, there is another famous conjecture in number theory which is closely connected with the theory of p-integral normal bases; to wit, Vandiver's conjecture, which states that the class number of $\mathbb{Q}(\zeta_p)^+$ is never divisible by p; p being any prime, and ζ_p a primitive p-th root of 1. It has been shown by Kersten and Michaliček (1985) that Vandiver's conjecture is true for p iff $\mathrm{P}(R, C_p) = 0$, R = ring of p-integers in $\mathbb{Q}(\zeta_p)$. The proof uses a kind of Spiegelungsprinzip (reflection technique). The same authors also proved a more general and more difficult) theorem, for which we refer to Kersten and Michaliček (1989b).

Some results of the aforementioned paper, and Thm. IV 2.2, have been reproved recently by Fleckinger and Nguyen Quang Do (1991). Their proofs are considerably shorter, but use stronger tools from Iwasawa theory.

Since there is a big problem with the descent for $p = 2$, this case has been excluded in Chap. IV (and most of the quoted papers as well). It is easy to see that everything still works for $p = 2$ as long as all groups Γ_n are still cyclic, which happens for instance if K contains $\sqrt{-1}$. The author has shown that at least Thm. IV 2.2 remains correct for $p = 2$ [Greither (1991b)]. To prove this, one needs a stronger version of IV 2.2 in the "good case" $\sqrt{-1} \in K$. To be precise, one proves a Δ-equivariant version of that theorem, with $\Delta = \mathrm{Gal}(K/\mathbb{Q})$, and $K/\mathbb{Q}$ supposed normal with $\sqrt{-1} \in K$. Then a sort of final descent is possible, and one may throw out the normality hypothesis as well as $\sqrt{-1}$ at the same time.

Finally it should be pointed out that there is a paper by Janelidze (1982) which deserves to be more widely known. For one thing, it contains a beautiful formula expressing $\mathrm{NB}(R, C_{p^n})$ in the form $\widetilde{\mathrm{Ext}}_{\mathbb{Z}\Gamma_n}(\mu_{p^n}, S_n^*)$ (adapted notation), where $\widetilde{\mathrm{Ext}}_{\mathbb{Z}\Gamma_n}(A, B)$ is the subgroup of $\mathrm{Ext}_{\mathbb{Z}\Gamma_n}(A, B)$ made up by all extensions $0 \to B \to E \to A \to 0$ such that $E \to A$ has a section as map of Γ_n-*sets*. (This peculiar $\widetilde{\mathrm{Ext}}$ is the Ext in a certain topos.) It is remarkable that Janelidze's and our results can be transformed into each other comparatively easily, while at the same time his techniques are markedly different form ours (he uses Hopf algebras and a good deal of category theory). Moreover, Janelidze also conjectured the "dual" formula $\mathrm{P}(R, C_{p^n}) \cong \mathrm{Hom}_{\mathbb{Z}\Gamma_n}(\mu_{p^n}, S_n^*)$ [Janelidze (1991)], which turned out to be equivalent to our description I 2.7 and I 3.6.

CHAPTER V

Geometric theory: cyclic extensions of finitely generated fields

In this chapter, we begin by explaining how our results on Galois extensions of (commutative) rings read in the language of varieties or schemes. We shall not use these extended results much, in particular non-affine schemes play no important part. Then we discuss C_{p^n}-extensions and $\mathbb{Z}_p$-extensions of fields K which are finitely generated field extensions of their prime field $\mathbb{Q}$. We shall obtain finiteness results for the group of C_{p^n}-extensions "unramified outside a given divisor" by a method different from that of Katz and Lang (1981). It is also proved that, for k the algebraic closure of $\mathbb{Q}$ in such a field K, k is an algebraic number field, and all $\mathbb{Z}_p$-extensions of K already come from $\mathbb{Z}_p$-extensions of k. This means that roughly speaking, the theory of $\mathbb{Z}_p$-extensions of K is already concentrated in number theory; the "geometry", i.e. passing form k to K, adds nothing new.

It is assumed in this chapter that the reader is reasonably familiar with the theory of varieties over an arbitrary ground field. The notion of étale extension will make an appearance for the sake of completeness, but we shall not use it seriously.

Let $p \neq 2$ throughout this chapter.

§1 Geometric prerequisites

Let k be a field. As one knows, the category of affine k-varieties is dual to the category of k-algebras of finite type without zero divisors. The category of affine reduced k-schemes of finite type is dual to the category of reduced k-algebras of finite type. (An algebra is *reduced* if it has no nonzero nilpotents.) An irreducible and reduced k-scheme of finite type is "the same" as a k-variety. We assume that all k-schemes are reduced and of finite type in this chapter.

A morphism of k-schemes f: $X \to Y$ is *finite* if $\mathcal{O}_X$ is finitely generated as an $f_*\mathcal{O}_Y$-module. One knows that finite morphisms are *surjective*, and *affine* (i.e.: Y has a covering by affine open sets Y_i such that all $f^{-1}(Y_i)$ are again affine.

Suppose now that a finite group G acts on the k-scheme X, and f: $X \to Y$ is a morphism of k-schemes such that "G acts over Y", i.e. $f = f\circ(\text{action of } \sigma)$ for all $\sigma \in G$.

Definition: In this situation, X/Y is called a *G-Galois covering* if: f is finite and étale (étale meaning flat and unramified), and if Y admits an open affine covering $Y = \bigcup Y_\iota$ such that with $X_\iota = f^{-1}(Y_\iota)$, the ring $\Gamma(X_\iota) = \mathcal{O}_X(X_\iota)$ becomes via $f^{\#}$ a G-Galois extension of $\Gamma(Y_\iota)$. Note that X_ι and Y_ι are both affine, so $X_\iota = \mathrm{Spec}(\Gamma(X_\iota))$, and $Y_\iota = \mathrm{Spec}(\Gamma(Y_\iota))$. Since the notion of G-Galois extension is stable under localization, standard arguments show that this definition is independent of the choice of the affine covering of Y.

Note the following consequence of this definition: if f: $X \to Y$ is G-Galois, then X is the union of G-stable open affine subsets. Hence the whole theory is essentially affine in nature, and the globalizaton to schemes is essentially trivial. Similarly as in the affine case, we may define sets $\mathrm{H}(Y,G)$ for every k-scheme Y (reduced and of finite type!) and every finite group G. If G is abelian, one may globalize the construction of **0**§3 and define a structure of abelian group on $\mathrm{H}(Y,G)$. This theory is not new, cf. SGA I. One can now rewrite Kummer theory and large parts of Chap. I for schemes, but we will not do so here. We need the following result in the sequel:

Proposition 1.1. *Let R be a k-algebra of finite type which is an integrally closed domain, and $K = \mathrm{Quot}(R)$. Let G be a finite group.*

a) If S/R is G-Galois, then S embeds into $K\otimes_R S$, and S is integrally closed in $K\otimes_R S$.

b) The canonical map $\mathrm{H}(R,G) \longrightarrow \mathrm{H}(S,G)$ is injective.

c) If S/R is G-Galois, G abelian, and S connected, then S is even a domain.

d) Let now be k a field, Y a normal k-variety with function field $k(Y)$, and Y' in Y open. Then the canonical map $\mathrm{H}(Y,G) \longrightarrow \mathrm{H}(Y',G)$, $(X \xrightarrow{f} Y) \mapsto (f^{-1}(Y') \xrightarrow{f} Y')$, is injective.

Proof. a) and b): The injectivity of $S \to K\otimes_R S$ follows from the fact that S is projective over R. The rest is just a restatement of Cor. **0** 4.2. (Cf. EGA IV, 6.5.4.)

c) Suppose S is connected. We first show that $K\otimes_R S$ is also connected. If $e \in K\otimes_R S$ is idempotent, then e is integral over R (since $e^2 = e$), and by a) we have $e \in S$, i.e. e is 0 or 1. Now $K\otimes_R S$ is artinian (as a finite extension of K). Since it is Galois over K, it is also reduced. On the other hand, it is connected. Therefore it must be a field, and $S \subset K\otimes_R S$ is a domain.

d) Consider the maps $\mathrm{H}(Y,G) \to \mathrm{H}(Y',G) \to \mathrm{H}(k(Y'),G)$, the last map being given by taking the generic fiber of $X \to Y'$. Note in this context that $k(Y')$ equals $k(Y)$. It suffices to show that the composite of the two maps is injective.

If Y is affine, then this follows directly from b). In general, we know at least: if $(f\colon X\to Y)$ goes to the trivial element of $\mathrm{H}(k(Y),G)$, then for each open affine $U \subset Y$, the restriction $f^{-1}(U) \to U$ is the trivial element of $\mathrm{H}(U,G)$. This (surprisingly) suffices to show the triviality of $X \to Y$: For each open subset $V \subset Y$, and $Z \to V$ the trivial G-Galois covering, i.e. $Z = Y\times G$ is a direct union of $|G|$ copies of Y, the group of G-automorphisms of Z over V is just G, since V is connected (even irreducible). Nota bene: this argument would fail in a general topological context. Hence: If $X' \to Y'$ is G-Galois, $Y' = U_1 \cup U_2$ (U_i open affine), and $(f^{-1}(U_i) \to U_i)$ is trivial for $i = 1,2$, then $X' \to Y'$ is itself already trivial. (Proof: Given U_i-isomorphisms $\alpha_i\colon f^{-1}(U_i) \cong U_i\times G$ for $i = 1,2$, we may look at the restrictions of α_i over $U_1 \cap U_2$; by the above, these restrictions differ only by some element σ of G, so if we repair this by replacing (say) α_1 through $\sigma\alpha_1$, we may glue together α_1 and α_2 to a global trivialization.) We now obtain by induction over the number of affine pieces needed to cover Y that $X \to Y$ is the trivial G-Galois covering. Q.E.D.

We have to review briefly Picard schemes for later use. For every smooth projective variety X over a field k, there exists an abelian k-group scheme $\mathbf{Pic}_X$ such that for all field extensions ℓ/k, we have $\mathrm{Pic}(X\otimes_k \ell) \cong \mathbf{Pic}_X(\ell)$ functorially. (There is much more to say about $\mathbf{Pic}_X$.) The connected component of the neutral element in $\mathbf{Pic}_X$ is written $\mathbf{Pic}_X^{(0)}$; it is an abelian variety. The quotient $\mathbf{Pic}_X/\mathbf{Pic}_X^{(0)}$ is "essentially" a finitely generated abelian group; in particular, for any field extension ℓ/k, $\mathbf{Pic}_X(\ell)/\mathbf{Pic}_X^{(0)}(\ell)$ is finitely generated. (More precisely, $\mathbf{Pic}_X/\mathbf{Pic}_X^{(0)}$ becomes a constant group scheme over the algebraic closure of k, and the underlying abelian group is finitely generated.) See Grothendieck (FGA), exposés 232 and 236 (1961/62), and Lang (1959). We will need the following result:

Theorem 1.2. *Let A be an abelian variety over a number field k. Let p ($\neq 2$) be a prime number, k_a an algebraic closure of k, and $\Omega_k = \mathrm{Aut}(k_a/k)$. Then the group*

$A(k_a)[p^\infty](-1)^{\Omega_k}$ is finite. (The affix $[p^\infty]$ means: take the p-primary torsion. The twist (-1) is the same as in Chap. I, defined via the canonical map ω: $\Omega_k \to \mathbb{Z}_p^*$, $\sigma(\zeta) = \zeta^{\omega(\sigma)}$ for $\zeta \in \mu_{p^\infty}(k_a)$. Note that Ω_k operates on $A(k_a)$ in a natural fashion.)

Proof. Let $k_\infty = k(\mu_{p^\infty})$, $\Omega' = \mathrm{Aut}(k_a/k_\infty)$. Then the operation of Ω' on the group $A(k_a)[p^\infty](-1)$ is untwisted, i.e. the fixed group $A(k_a)[p^\infty](-1)^{\Omega_k}$ is contained in $A(k_a)[p^\infty]^{\Omega'} = A(k_a^{\Omega'})[p^\infty] = A(k_\infty)[p^\infty]$. By a theorem of Imai (1975) and Serre (1974), cf. also Ribet (1982), the group $A(k_\infty)_{tors}$ is already finite, which proves the theorem.

Addendum: The proof of the theorem quoted above is quite brief in Serre (1974) and Ribet(1982), and in Imai(1975) there is a restrictive extra hypothesis. For this reason, we indicate another proof of Thm. 1.2. We use the Tate module $T_p(A) = \varprojlim A[p^n]$ and $V_p(A) = \mathbb{Q}_p \otimes_{\mathbb{Z}_p} T_p(A)$. Furthermore we use the dual abelian variety A^ and the Weil pairing. (For all these matters, see Milne (1986b).) We have the following implications, with A_p being short for $A[p^\infty]$, and $\Omega = \Omega_k$:

$$A_p(-1)^\Omega \text{ infinite} \Rightarrow \mathrm{Hom}(\mathbb{Q}_p/\mathbb{Z}_p, A_p(-1))^\Omega = T_p(A)(-1)^\Omega \text{ infinite}$$
$$\Rightarrow V_p(A)(-1)^\Omega \text{ infinite.}$$

Now there is a canonical nondegenerate pairing $V_p(A) \times V_p(A\char`^) \to \mathbb{Q}_p(1)$, hence $V_p(A)(-1) \cong \mathrm{Hom}(V_p(A\char`^), \mathbb{Q}_p)$, and we can continue:

$$\ldots \Rightarrow \mathrm{Hom}(V_p(A\char`^), \mathbb{Q}_p)^\Omega \text{ infinite, in particular} \neq 0$$
$$\Rightarrow V_p(A\char`^)_\Omega \text{ nonzero.}$$

(The subscript Ω means: the greatest factor module on which Ω acts trivially.) Now $V_p(A\char`^)$ is a semisimple $\mathbb{Q}_p\Omega$-module, by the validity of the Tate conjectures over a number field (Faltings (1983)). Therefore $V_p(A\char`^)$ has a nonzero Ω-trivial factor module iff it has a nonzero Ω-trivial submodule, hence we arrive at the statement $V_p(A\char`^)^\Omega \neq 0$, i.e. infinite, which implies that $(A\char`^)_p(k)$ is infinite, a contradiction to the Mordell-Weil theorem which says that $A\char`^(k)$ is finitely generated, q.e.d.

Corollary 1.3. *Let $\Gamma = \mathrm{Aut}(k_\infty/k)$ with $k_\infty = k(\mu_{p^\infty})$, and X a smooth projective k-variety. Write X_∞ for $X \otimes_k k_\infty$. Then the p-primary torsion of* $\mathrm{Pic}(X_\infty)(-1)^\Gamma$ *is finite.*

Proof. By 1.2, and since $A = \mathbf{Pic}_X^{(0)}$ is an abelian variety, $\mathbf{Pic}_X^{(0)}(k_\infty)(-1)^\Gamma$ is finite. Let $N = \mathbf{Pic}_X(k_\infty)/\mathbf{Pic}_X^{(0)}(k_\infty)$. It suffices to show that $N(-1)^\Gamma$ is finite. Now every element of N already comes from some $\mathrm{Pic}(X_n)$ ($X_n = X \otimes k(\zeta_n)$, ζ_n a primitive p^n-th root of 1), and N is finitely generated. Hence Γ operates on N via a finite quotient, and it is easy to see that this implies $N(-1)^\Gamma$ finite (cf. III 1.4).

Corollary 1.4. *The result of Cor. 1.3 holds also for all smooth quasi-projective k-varieties X.*

Proof. Embed X in a smooth projective k-variety X'. (Note that we need resolution of singularities in char. zero to do this.) Let $D = X'_\infty - X_\infty$, and let $D_1, \ldots, D_s$ be the components of D with codimension one in X'_∞. Then $\mathrm{Pic}(X_\infty) = \mathrm{Pic}(X'_\infty)/U$ with U the subgroup generated by the divisor classes $[D_1], \ldots, [D_s]$. We get a sequence

$$\mathrm{Pic}(X'_\infty)(-1)^\Gamma \longrightarrow \mathrm{Pic}(X_\infty)(-1)^\Gamma \longrightarrow \mathrm{H}^1(\Gamma, U(-1)).$$

By the same argument as before, there is a $\gamma \in \Gamma$, $\gamma \neq 1$, which operates trivially on U, hence as multiplication with a p-adic integer $\neq 1$ on $U(-1)$. It follows that $U(-1)/(1-\gamma)\cdot U(-1)$ is finite. From this one easily gets that $\mathrm{H}^1(\Gamma, U(-1))$ is finite. The assertion of 1.4 now follows from 1.3.

§2 $\mathbb{Z}_p$-extensions of absolutely finitely generated fields

We only consider fields of characteristic zero. Such a field K is called *absolutely finitely generated* (*afg* for short) if it is a function field over $\mathbb{Q}$, i.e. if it can be written $K = \mathbb{Q}(x_1, \ldots, x_n)$, $x_i \in K$. We then have $\mathrm{tr.deg}_{\mathbb{Q}}(K) \leq n$.

Standing **notation**: For an *afg* field K, k always denotes the algebraic closure of $\mathbb{Q}$ in K.

Lemma 2.1. *a) Any subfield of an* afg *field is itself an* afg *field.*
b) If K is afg, *then k is an algebraic number field.*

Proof. a) $\Rightarrow$ b) holds since an *afg* field which is algebraic over $\mathbb{Q}$ is obviously finite over $\mathbb{Q}$.

a) This it not new, but not too widely known either. We argue as follows: For any field L, define an L-algebra A of essentially finite type (*eft* for short) to be an L-algebra which is a localization of another L-algebra A_0 of finite type. It follows that such an A never contains infinitely many idempotents since A is noetherian. If K is afg, $K' \subset K$ a subfield, and L a field extension of K', then $L \otimes_{K'} K$ is *eft* over L. Let now $E \subset K$ be any subfield, and choose $K' \subset E$ with E/K' algebraic and $K'/\mathbb{Q}$ purely transcendental. We have to show that E is in fact finite over K'. Let L be an algebraic closure of K'. Then $L \otimes_{K'} K$ is *eft* over L and contains $L \otimes_{K'} E$. This latter algebra would contain infinitely many idempotents if $[E:L]$ were infinity, contradicting the fact that no *eft* algebra contains infinitely many idempotents.

In the next lemma, we only need that k is relatively algebraically closed in K.

Lemma 2.2. *a) For every algebraic extension ℓ/k, the algebra $\ell \otimes_k K$ is also a field.*
b) For any abelian group G, the natural map $H(k,G) \longrightarrow H(K,G)$ is injective.

Proof. a) This is well-known. (Sketch: Assume char$(k) = 0$ for simplicity. We may then assume $\ell = k(\alpha)$. Let f be the minimal polynomial of α over k. If f had a nontrivial divisor g in $K[X]$, then the coefficients of g would be algebraic over k, and not all in k, contradiction.)

b) This follows easily from a). Q.E.D.

It turns out that it is very practical to consider "C_{p^n}-extensions with variable n" in the sequel instead of $\mathbb{Z}_p$-extensions. More precisely:

Definition and Lemma 2.3. *For a connected ring R, let*

$$H(R,\mathbb{Q}_p/\mathbb{Z}_p) = \varinjlim H(R,C_{p^n}),$$

where the inductive limit is taken along the canonical injections $C_{p^n} \to C_{p^{n+1}}$ which map σ_n to σ_{n+1}^p. Since $H(R,-) \cong \mathrm{Hom}_{cont}(\Psi_R,-)$, we have

$$\begin{aligned} H(R,\mathbb{Q}_p/\mathbb{Z}_p) &\cong \varinjlim \mathrm{Hom}_{cont}(\Psi_R,C_{p^n}) \\ &= \mathrm{Hom}_{cont}(\Psi_R,\mathbb{Q}_p/\mathbb{Z}_p), \end{aligned}$$

because $\varinjlim C_{p^n} = \mathbb{Q}_p/\mathbb{Z}_p$ with the discrete topology, and Ψ_R is compact. Moreover, we have for all $\nu \geq 1$:

$$H(R,C_{p^\nu}) \cong H(R,\mathbb{Q}_p/\mathbb{Z}_p)[p^\nu].$$

Let now E be any field of char. $\neq p$ (p an odd prime), $E_n = E(\zeta_n)$ (where as always ζ_n is a primitive p^n-th root of unity), $E_\infty = \bigcup E_n$. Recall that $n_0(E) = n_0$ was defined as the maximal ν with $\zeta_\nu \in E_1$ (maybe $n_0 = \infty$). In I §3 we obtained for $n \geq n_0$ an exact sequence

$$(*) \quad 0 \longrightarrow \mathrm{Ker}(j_n) \longrightarrow H(E,C_{p^n}) \longrightarrow \left(E_n^*/p^n\right)(-1)^{\Gamma_n} \longrightarrow \mu_{p^n}/\mu_{p^{n_0}} \longrightarrow 1.$$

Here $\mathrm{Ker}(j_n)$ is formed by cyclotomic extensions, and has order p^{n-n_0}, hence $\mathrm{Ker}(j_n)$ is the subgroup of order p^{n-n_0} in the group $\langle A_n(E)\rangle$ generated by the p^n-th cyclotomic p-extension of E. We now pass to the inductive limit in the sequence $(*)$. Note first of all that $\varinjlim$ is an exact functor (in contrast to $\varprojlim$). We obtain:

$\varinjlim H(E,C_{p^n}) = H(E,\mathbb{Q}_p/\mathbb{Z}_p)$;

$\varinjlim \mathrm{Ker}(j_n)$ is the subgroup of $H(E,\mathbb{Q}_p/\mathbb{Z}_p)$ consisting of *all* cyclotomic p-extensions of E; this can be (noncanonically) identified with $\mathbb{Q}_p/\mathbb{Z}_p$;

similarly,

$\varinjlim \mu_{p^n}/\mu_{p^{n_0}} = \mu_{p^\infty}/\mu_{p^{n_0}} \cong \mathbb{Q}_p/\mathbb{Z}_p(1)$;

and finally we claim that

$$\lim_{\to} \left(E_n^*/p^n\right)(-1)^{\Gamma_n} \cong \left(\mathbb{Q}_p/\mathbb{Z}_p \otimes_{\mathbb{Z}} E_\infty^*\right)(-1)^{\Gamma}. \qquad (\Gamma = \mathrm{Aut}(E_\infty/E))$$

(Proof of claim: First one checks using Kummer theory that the map $E_n^*/p^n \to E_{n+1}^*/p^{n+1}$ in question is the p-th power map. Hence we do get a well-defined map α: $\lim_{\to}\left(E_n^*/p^n\right) = \lim_{\to}\left(\mathbb{Z}/p^n \otimes E_n^*\right) \longrightarrow \lim_{\to}(\mathbb{Z}/p^n) \otimes \lim_{\to} E_n^* = \mathbb{Q}_p/\mathbb{Z}_p \otimes E_\infty^*$. It is easy to see that α is an isomorphism, and we still have an isomorphism after twisting and taking Γ-invariants (note Γ_n is a factor group of Γ).) We obtain therefore:

Proposition 2.4. *With E, E_∞ and Γ as above, there is an exact sequence*

$$0 \longrightarrow \mathbb{Q}_p/\mathbb{Z}_p \longrightarrow H(E,\mathbb{Q}_p/\mathbb{Z}_p) \to \left(\mathbb{Q}_p/\mathbb{Z}_p \otimes_{\mathbb{Z}} E_\infty^*\right)(-1) \longrightarrow \mathbb{Q}_p/\mathbb{Z}_p(-1) \to 0.$$

Let us now return to the situation: K an *afg* field, k the algebraic closure of $\mathbb{Q}$ inside K. Then we have $n_0(K) = n_0(k)$ (use e.g. 2.2 a)), we may identify the Galois groups of k_∞ and K_∞ (2.2 a) again), and we obtain a commutative diagram:

$$\begin{array}{ccccccccc} 0 \to & \mathbb{Q}_p/\mathbb{Z}_p & \to & H(k,\mathbb{Q}_p/\mathbb{Z}_p) & \to & \left(\mathbb{Q}_p/\mathbb{Z}_p \otimes_{\mathbb{Z}} k_\infty^*\right)(-1)^{\Gamma} & \to & \mathbb{Q}_p/\mathbb{Z}_p(-1) \to 0 \\ & \| & & \downarrow \beta & & \downarrow \gamma & & \| \\ 0 \to & \mathbb{Q}_p/\mathbb{Z}_p & \to & H(K,\mathbb{Q}_p/\mathbb{Z}_p) & \to & \left(\mathbb{Q}_p/\mathbb{Z}_p \otimes_{\mathbb{Z}} K_\infty^*\right)(-1)^{\Gamma} & \to & \mathbb{Q}_p/\mathbb{Z}_p(-1) \to 0, \end{array}$$

where β and γ are induced by the inclusions $k \subset K$ and $k_\infty \subset K_\infty$.

Lemma 2.5. *The canonical maps β and γ are injective, and*

$$\mathrm{Coker}(\gamma) \cong \left(\mathbb{Q}_p/\mathbb{Z}_p \otimes_{\mathbb{Z}} (K_\infty^*/k_\infty^*)\right)(-1)^{\Gamma}.$$

Proof. The statement for β follows from Lemma 2.2 b).

We know $k_\infty \otimes_k K = K_\infty$. By Galois descent, one easily shows that k_∞ is again relatively algebraically closed in K_∞. In particular K_∞^*/k_∞^* has no torsion, hence

$$0 \longrightarrow \mathbb{Q}_p/\mathbb{Z}_p \otimes_{\mathbb{Z}} k_\infty^* \longrightarrow \mathbb{Q}_p/\mathbb{Z}_p \otimes_{\mathbb{Z}} K_\infty^* \longrightarrow \mathbb{Q}_p/\mathbb{Z}_p \otimes_{\mathbb{Z}} (K_\infty^*/k_\infty^*) \longrightarrow 0$$

is exact. This gives at once that γ is injective. The sequence remains exact on twisting and taking Γ-invariants, by Cor. II 2.2. (A little care is necessary to pass to the limit.) This gives the claimed information concerning $\mathrm{Coker}(\gamma)$.

From this lemma and the preceding diagram, one obtains by a routine diagram chase (using the snake lemma):

Proposition 2.6. *With the above notation, one has an isomorphism*

$$\frac{H(K,\mathbb{Q}_p/\mathbb{Z}_p)}{H(k,\mathbb{Q}_p/\mathbb{Z}_p)} \cong \left(\mathbb{Q}_p/\mathbb{Z}_p \otimes (K_\infty^*/k_\infty^*)\right)(-1)^{\Gamma}.$$

This is the starting point of the calculation. We shall now use some facts from algebraic geometry. A good reference is Hartshorne (1976). Recall: K is absolutely finitely generated over $\mathbb{Q}$, and k (the algebraic closure of $\mathbb{Q}$ in K) is a number field. Since one disposes of resolution of singularities in char. zero, there exists a smooth projective k-variety V with function field $k(V)$ isomorphic to K. (For the case $\mathrm{tr.deg}_k K \le 2$, this has been known for a long time. In general, one has to use Hironaka's theorem.) The ring of global sections $\Gamma(V,\mathcal{O}_V) = \mathcal{O}(V)$ is then k (since it is finite over k). Hence $\mathcal{O}(V)^* = k^*$.

Let $\mathrm{Div}(V)$ be the divisor group of V. $\mathrm{Div}(V)$ is free abelian on the set of (irreducible) 1-codimensional subvarieties of V. The group of principal divisors $\mathrm{Prin}(V)$ is isomorphic to $K^*/\mathcal{O}(V)^* = K^*/k^*$. (A function $f \in K^*$ has trivial divisor iff it is invertible on V.) Since $\mathrm{Div}(V)/\mathrm{Prin}(V) \cong \mathrm{Pic}(V)$ by the smoothness of V, we obtain a short exact sequence:

$$(*) \quad 1 \longrightarrow K^*/k^* \longrightarrow \mathrm{Div}(V) \longrightarrow \mathrm{Pic}(V) \longrightarrow 0.$$

An analogous sequence exists for k_∞, K_∞, and $V_\infty = V \otimes_k k_\infty$ in the place of k, K, and V (one again uses that k_∞ is algebraically closed in K_∞):

$$(**) \quad 1 \longrightarrow K_\infty^*/k_\infty^* \longrightarrow \mathrm{Div}(V_\infty) \longrightarrow \mathrm{Pic}(V_\infty) \longrightarrow 0.$$

We now apply $\mathbb{Q}_p/\mathbb{Z}_p \otimes_{\mathbb{Z}} -$ to (**) and use that $\mathrm{Tor}_{\mathbb{Z}}(\mathbb{Q}_p/\mathbb{Z}_p, M)$ is just the p-primary torsion part of an abelian group M, and that $\mathrm{Div}(V_\infty)$ has no torsion:

$$(3) \quad 0 \longrightarrow \mathrm{Pic}(V_\infty)[p^\infty] \longrightarrow (\mathbb{Q}_p/\mathbb{Z}_p) \otimes (K_\infty^*/k_\infty^*) \longrightarrow (\mathbb{Q}_p/\mathbb{Z}_p) \otimes \mathrm{Div}(V_\infty).$$

All groups involved are $\Gamma = \mathrm{Aut}(k_\infty/k)$-modules in a canonical fashion, so we obtain by twisting and taking Γ-invariants:

$$(4) \quad 0 \longrightarrow \mathrm{Pic}(V_\infty)[p^\infty](-1)^\Gamma \longrightarrow \big((\mathbb{Q}_p/\mathbb{Z}_p) \otimes (K_\infty^*/k_\infty^*)\big)(-1)^\Gamma \longrightarrow X,$$

with $X = \big((\mathbb{Q}_p/\mathbb{Z}_p) \otimes \mathrm{Div}(V_\infty)\big)(-1)^\Gamma$.

Lemma 2.7. *X is a direct sum of cyclic torsion groups.*

Proof. Let $\mathrm{P}(V_\infty)$ be the set of prime divisors (i.e. of 1-codimensional subvarieties) of V_∞. Then $\mathrm{Div}(V_\infty)$ is the free abelian group over $\mathrm{P}(V_\infty)$. Now every $Q \in \mathrm{P}(V_\infty)$ is already defined over some field $k_n \subset k_\infty$, and consequently invariant under $\Gamma_n = \mathrm{Aut}(k_\infty/k_n) \subset \Gamma$. Therefore the Γ-module $\mathrm{Div}(V_\infty)$ is a direct sum of permutation modules, i.e. of modules isomorphic to $\mathbb{Z}[\Gamma/\Gamma']$, with $\Gamma' \subset \Gamma$ an open subgroup. With the same argument as in Lemma III 1.4, one finds that $\big(\mathbb{Q}_p/\mathbb{Z}_p \otimes \mathbb{Z}[\Gamma/\Gamma']\big)(-1)^\Gamma$ is finite (even cyclic), of order at most p^m if $\Gamma_m \subset \Gamma'$. (Again, the quoted lemma only treats $\big((\mathbb{Z}/p^n) \otimes \mathbb{Z}[\Gamma/\Gamma']\big)(-1)^\Gamma$, but it is no problem to pass to the direct limit.) This proves the lemma.

We now are able to prove:

Theorem 2.8. *The group* $\mathrm{H}(K,\mathbb{Q}_p/\mathbb{Z}_p)/\mathrm{H}(k,\mathbb{Q}_p/\mathbb{Z}_p)$ *is reduced, i.e.: it contains no nonzero divisible subgroup.*

Proof. By 2.5, the group in question is isomorphic to $\left(\mathbb{Q}_p/\mathbb{Z}_p \otimes (K_\infty^*/k_\infty^*)\right)(-1)^\Gamma$. Since we have the exact sequence (4), it is visibly enough to show that neither X nor $\mathrm{Pic}(V_\infty)[p^\infty](-1)^\Gamma$ contain a nonzero divisible subgroup. For X, this follows from 2.7 since any homomorphism from a divisible group to a cyclic group is zero. For the Pic term, Cor. 1.3 gives even finiteness, which is much more than we need. Q.E.D.

As a consequence we get one of the main results of this chapter:

Theorem 2.9. *For any field K which is finitely generated as a field extension of* $\mathbb{Q}$, *and k the algebraic closure of* $\mathbb{Q}$ *in K, all* $\mathbb{Z}_p$*-extensions of K already come from k, i.e. the canonical map*

$$\alpha:\quad \mathrm{H}(k,\mathbb{Z}_p) \longrightarrow \mathrm{H}(K,\mathbb{Z}_p)$$

is an isomorphism.

Proof. By Lemma 2.2, the map $\mathrm{H}(k,G) \longrightarrow \mathrm{H}(K,G)$ is injective for all finite abelian groups G. By general abstract nonsense, it follows that the abelianized absolute Galois group Ω_k is canonically isomorphic to a factor group of Ω_K (see **0**, Thm. 3.5). Let $\Theta = \mathrm{Ker}(\Omega_K \longrightarrow \Omega_k)$. Then a homomorphism $f\colon \Omega_K \to \mathbb{Z}_p$ is in $\mathrm{Im}(\alpha)$ iff f kills Θ. Now we start with any nontrivial $\mathbb{Z}_p$-extension of K, described by some $0 \neq f\colon \Omega_K \to \mathbb{Z}_p$. Then the maps $f_n\colon \Omega_K \to \mathbb{Z}_p \to \mathbb{Z}/p^n\mathbb{Z}$ $(n \in \mathbb{N})$ give a submodule $W = \langle f_n \mid n \in \mathbb{N}\rangle$ of $\mathrm{Hom}_{cont}(\Omega_K,\mathbb{Q}_p/\mathbb{Z}_p)$, and W is itself isomorphic to $\mathbb{Q}_p/\mathbb{Z}_p$. By Thm. 2.8, W lies entirely in $\mathrm{Hom}_{cont}(\Omega_k,\mathbb{Q}_p/\mathbb{Z}_p) = \mathrm{H}(k,\mathbb{Q}_p/\mathbb{Z}_p)$, hence all f_n annihilate Θ. Therefore also f annihilates Θ, hence $f \in \mathrm{Im}(\alpha)$, q.e.d.

Remark. There is the following way to think about Thm. 2.9. It seems reasonable to call $\mathbb{Z}_p$-extensions of K "number-theoretic" if they come from $\mathbb{Z}_p$-extensions of k, and to call the factor group $\mathrm{H}(K,\mathbb{Z}_p)/\mathrm{H}(k,\mathbb{Z}_p)$ the group of "geometric" $\mathbb{Z}_p$-extensions. Cf. Katz and Lang (1981). The result 2.9 then says that the latter group is zero.

§ 3 A finiteness result

In this section we, by and large, reprove a finiteness theorem due to Katz and Lang (1981). The methods in that paper heavily use "geometric class field theory" (involving generalized Jacobians), while our method, as the reader has seen already in §2, is Kummer theory plus descent; some use of Picard varieties is still necessary. It is likely that our result 2.9 can also be proved with the methods of Katz and Lang.

Let k be a number field and $V = \mathrm{Spec}(A)$ a geometrically connected, smooth affine k-variety. (This means that $k_a \otimes_k A$ is a regular domain, and A is of finite type over k. In particular, k is algebraically closed in $\mathrm{Quot}(A)$, otherwise $k_a \otimes_k A$ would have nontrivial idempotents.) Let $K = \mathrm{Quot}(A) = k(V)$. The result is as follows:

Theorem 3.1. *a) The image of* $\mathrm{H}(k, \mathbb{Q}_p/\mathbb{Z}_p)$ *has finite index in* $\mathrm{H}(A, \mathbb{Q}_p/\mathbb{Z}_p)$.

b) For any $n \in \mathbb{N}$, *the image of* $\mathrm{H}(k, C_{p^n})$ *in* $\mathrm{H}(A, C_{p^n})$ *has finite index.*

Remarks. a) The map $\mathrm{H}(k, \mathbb{Q}_p/\mathbb{Z}_p) \longrightarrow \mathrm{H}(A, \mathbb{Q}_p/\mathbb{Z}_p)$ is injective, since already the composite map $\mathrm{H}(k, \mathbb{Q}_p/\mathbb{Z}_p) \longrightarrow \mathrm{H}(K, \mathbb{Q}_p/\mathbb{Z}_p)$ is injective by 2.2 b).

b) If V has a k-rational point, i.e. A has an augmentation $\varepsilon\colon A \twoheadrightarrow k$, then the theorem may be restated by saying that $\mathrm{Ker}(\mathrm{H}(\varepsilon, \mathbb{Q}_p/\mathbb{Z}_p))$ is finite.

Proof of **3.1**. We first note that b) easily follows from a) if one uses $\mathrm{H}(k, C_{p^n}) =$ p^n-torsion of $\mathrm{H}(k, \mathbb{Q}_p/\mathbb{Z}_p)$, and the corresponding equality for A.

Let $k_\infty = k(\mu_{p^\infty})$ as usual, $A_\infty = k_\infty \otimes_k A$. As in §2 (following 2.4), we have exact sequences

$$\begin{array}{ccccccccc} 0 \longrightarrow & \mathbb{Q}_p/\mathbb{Z}_p & \longrightarrow & \mathrm{H}(k, \mathbb{Q}_p/\mathbb{Z}_p) & \longrightarrow & \mathrm{H}(k_\infty, \mathbb{Q}_p/\mathbb{Z}_p)^\Gamma & \longrightarrow & \mathbb{Q}_p/\mathbb{Z}_p(1) & \longrightarrow 1 \\ & \| & & \beta \downarrow & & \gamma \downarrow & & \| & \\ 0 \longrightarrow & \mathbb{Q}_p/\mathbb{Z}_p & \longrightarrow & \mathrm{H}(A, \mathbb{Q}_p/\mathbb{Z}_p) & \longrightarrow & \mathrm{H}(A_\infty, \mathbb{Q}_p/\mathbb{Z}_p)^\Gamma & \longrightarrow & \mathbb{Q}_p/\mathbb{Z}_p(1) & \longrightarrow 1. \end{array}$$

We know already that β is injective, By the same token, γ is injective. If we knew that $\mathrm{H}^1(\Gamma, X) = 0$ with $X = \mathrm{H}(A_\infty, \mathbb{Q}_p/\mathbb{Z}_p)/\mathrm{H}(k_\infty, \mathbb{Q}_p/\mathbb{Z}_p)$, then we would have $\mathrm{Coker}(\gamma) = X^\Gamma$. Now by 1.1 b), X embeds into $\mathrm{H}(K_\infty, \mathbb{Q}_p/\mathbb{Z}_p)/\mathrm{H}(k_\infty, \mathbb{Q}_p/\mathbb{Z}_p)$ which equals $\mathbb{Q}_p/\mathbb{Z}_p \otimes (K_\infty^*/k_\infty^*)(-1)$ (proof of 2.5). Hence X can be written as the inductive limit of groups of the form $\mathbb{Q}_p/\mathbb{Z}_p \otimes M(-1)$, where M is a torsion-free Γ-module on which Γ operates vie a finite factor group Γ/Γ_n ($n \in \mathbb{N}$). By Cor. II 2.2, the cohomology $\mathrm{H}^1(\Gamma, \mathbb{Q}_p/\mathbb{Z}_p \otimes M(-1))$ is zero, hence also $\mathrm{H}^1(\Gamma, X) = 0$, and $X^\Gamma =$ $\mathrm{coker}(\gamma)$.

By the same diagram chase as the one leading to 2.6, we get

$$\frac{H(A,\mathbb{Q}_p/\mathbb{Z}_p)}{H(k,\mathbb{Q}_p/\mathbb{Z}_p)} = \left(\frac{H(A_\infty,\mathbb{Q}_p/\mathbb{Z}_p)}{H(k_\infty,\mathbb{Q}_p/\mathbb{Z}_p)}\right)^{\Gamma},$$

and it remains to calculate the right hand side. By Kummer theory and passage to the inductive limit one has, since A_∞ and k_∞ are p^n-kummerian for all $n \geq 1$:

$$\begin{array}{ccccccc}
 & & 0 & & 0 & & \\
 & & \downarrow & & \downarrow & & \\
 & & \mathbb{Q}_p/\mathbb{Z}_p \otimes k_\infty^* & = & H(k_\infty,\mathbb{Q}_p/\mathbb{Z}_p) & & \\
 & & \downarrow & & \downarrow & & \\
0 \longrightarrow & & \mathbb{Q}_p/\mathbb{Z}_p \otimes A_\infty^* & \longrightarrow & H(A_\infty,\mathbb{Q}_p/\mathbb{Z}_p) & \longrightarrow & \mathrm{Pic}(A_\infty)[p^\infty] \longrightarrow 0 \\
 & & \downarrow & & & & \\
 & & \mathbb{Q}_p/\mathbb{Z}_p \otimes (A_\infty^*/k_\infty^*) & & & & \\
 & & \downarrow & & & & \\
 & & 0, & & & &
\end{array}$$

hence a short exact sequence

$$(*) \quad 0 \longrightarrow \mathbb{Q}_p/\mathbb{Z}_p \otimes (A_\infty^*/k_\infty^*) \longrightarrow \frac{H(A_\infty,\mathbb{Q}_p/\mathbb{Z}_p)}{H(k_\infty,\mathbb{Q}_p/\mathbb{Z}_p)} \longrightarrow \mathrm{Pic}(A_\infty)[p^\infty] \longrightarrow 0.$$

We twist by ..(−1) and take Γ-invariants in (*). By Cor. 1.4, $\mathrm{Pic}(A_\infty)[p^\infty](-1)^\Gamma$ is finite. It therefore remains to show:

$$(**) \qquad \left(\mathbb{Q}_p/\mathbb{Z}_p \otimes (A_\infty^*/k_\infty^*)\right)(-1)^\Gamma \text{ is finite.}$$

By the theorem of Rosenlicht (Rosenlicht (1961)), the group $F = A_\infty^*/k_\infty^*$ is finitely generated and free. Pick $\eta_1, \ldots, \eta_s \in A_\infty^*$ which generate it mod k_∞^*. Then all η_i are already in some A_n (= $A[\zeta_n]$, ζ_n primitive p^n-th root of 1), hence Γ operates on A_∞^*/k_∞^* via some factor Γ/Γ_n. It is then easily seen that $\left(\mathbb{Q}_p/\mathbb{Z}_p \otimes F\right)(-1)^\Gamma$ has at most p^{ns} elements. (Take a topological generator γ of Γ_n; it operates trivially on F, hence it operates as multiplication by $\omega(\gamma)$ on $\left(\mathbb{Q}_p/\mathbb{Z}_p \otimes F\right)(-1)$. Now $\omega(\gamma) = 1 + p^n u$ with $u \in \mathbb{Z}_p$ a unit, hence $\left(\mathbb{Q}_p/\mathbb{Z}_p \otimes F\right)(-1)^{\gamma - 1} = \left(\mathbb{Q}_p/\mathbb{Z}_p \otimes F\right)[p^n]$). This proves (**), and the theorem follows.

Corollary 3.2. *The statement of Thm.* 3.1 *also holds for k-varieties V which are geometrically normal* (*not necessarily smooth*).

Proof. V contains an open affine smooth subvariety V', and Thm. 3.1 applies to V', i.e. $\mathrm{Coker}(H(k,\mathbb{Q}_p/\mathbb{Z}_p) \to H(V,\mathbb{Q}_p/\mathbb{Z}_p))$ is finite. On the other hand, the map $H(V,\mathbb{Q}_p/\mathbb{Z}_p) \longrightarrow H(V',\mathbb{Q}_p/\mathbb{Z}_p)$ is injective by 1.1 d); an easy argument now shows that the cokernel of $H(k,\mathbb{Q}_p/\mathbb{Z}_p) \to H(V,\mathbb{Q}_p/\mathbb{Z}_p)$ is also finite, q.e.d.

Remarks and Complements.

a) It goes without saying that the methods employed here do not at all diminish the importance of the geometric class field theory approach. Our aim has been to offer an alternative approach which perhaps provides some additional insight. Some difficult theorems were used: resolution of singularities, the existence of Picard varieties, and finiteness of the Neron–Severi group. On the other hand, we did not use generalized Jacobians, or the delicate specialization techniques from algebraic geometry ("elementary fibrations") as in the work of Katz and Lang.

b) Both Thm. 2.9 and 3.1 tell us that there are *no* or *only few* "geometric" extensions of an *afg* field K, or a geometrically normal variety V over a number field. The situation is totally different for varieties (function fields) over algebraically closed ground fields. This can be seen already by looking at coverings of an elliptic curve. In another direction, finiteness also breaks down if we allow non-normal varieties. The simplest example sems to be the α–curve V over any base field k, which is obtained by identifying the points 0 and 1 in the affine line. Thus V has one double point (with normal crossing). One can show that V has cyclic coverings of any degree which do not come from k. More precisely, one can show via appropriate Mayer–Vietoris sequences (Greither and Haggenmüller (1982)) that $\mathrm{H}(V,\mathbb{Q}_p/\mathbb{Z}_p) \cong \mathrm{H}(k,\mathbb{Q}_p/\mathbb{Z}_p) \oplus \mathbb{Q}_p/\mathbb{Z}_p$.

In this context, one can prove two other results for V a variety over a number field k:

(*i*) The factor group $\mathrm{H}(V,\mathbb{Z}_p)/\mathrm{H}(k,\mathbb{Z}_p)$ is finitely generated;

and

(*ii*) The group $\mathrm{NB}(V,\mathbb{Z}_p)$ consists precisely of all $\mathbb{Z}_p$–extensions of A which come from k.

Thus, in a sense, the $\mathbb{Z}_p$–extensions "which come from geometry" tend to fail to have normal bases. It is a pleasant exercise to check this for the α–curve.

CHAPTER VI

Cyclic Galois theory without the condition "$p^{-1} \in R$"

We give here a rather detailed description of C_{p^n}-extensions of a ring R which contains a primitive p^n-th root ζ_n of unity but not necessarily p^{-1}. Here p is a fixed prime, now allowed to be 2, and n is any exponent. Some main ideas in this theory go back to number-theoretical work of Hasse. In the last section, we also dismiss the condition "$\zeta_n \in R$", with a view towards descent theory and generic C_{p^n}-extensions. Some preliminaries are needed: Galois theory in characteristic p, Witt rings, and patching techniques.

§1 Witt rings, and Artin–Schreier theory for rings of characteristic p

This section gives a short presentation of Artin–Schreier theory which will be needed later. The results are the same as for *fields* of char. p in the classic of Witt (1936), and can be found e.g. in Wyler (1987) [Prop. 2.4, Cor. 2.6]. For the reader's convenience, we include proofs or at least indications of proof.

Let p be a fixed prime, R a ring of char. p (i.e. $p\cdot 1_R = 0$), and $q = p^n$ a power of p $(n \geq 1)$. One defines the *ring* $W_n(R)$ *of Witt vectors of length n over R* exactly as in the case R a field (Witt (1936)). We agree to denote Witt vectors by n-tuples or letters with underbar: $\underline{x} = (x_0, \ldots, x_{n-1})$, $\underline{y}$, . . ., and Witt addition (subtraction) by $\dotplus$ $(\dot{-})$. $W_n(R)$ has a canonical automorphism $(-)^P$ given by $(x_0,\ldots,x_{n-1})^P = (x_0^{\,P},\ldots,x_{n-1}^{\,P})$. Almost always, this is not the p-th power map in the ring $W\,(R)$.

Fix a generator σ of C_{p^n}. We define a map

$$\begin{aligned} \Phi:\quad W_n(R) &\longrightarrow \mathrm{H}(R, C_{p^n}), \\ \underline{x} &\longmapsto S_{\underline{x}} = R[\underline{Y}]/(\underline{Y}^P \dot{-} \underline{Y} \dot{-} \underline{x}), \text{ with } \sigma(\underline{Y}) = \underline{Y} \dotplus \mathbf{1}. \end{aligned}$$

Some explanations are in order: To adjoin $\underline{Y}$ just means adjoining n variables $Y_0, Y_1, \ldots, Y_{n-1}$. The relation $\underline{Y}^P \dot{-} \underline{Y} \dot{-} x$ has to be understood as an n-tuple of relations involving the Y_i and x_i, all of which are to be factored out. Similarly, the rule for σ is actually an n-tuple of defining rules. $\mathbf{1} = (1,0\ldots,0)$ is the unit element of $W_n(R)$.

Theorem 1.1. *For any ring R of char. p, the map* Φ *above induces an isomorphism*

$$W_n(R)/\{\underline{x}^P \dot{-} \underline{x} \mid \underline{x} \in W_n(R)\} \longrightarrow \mathrm{H}(R, C_{p^n}).$$

Proof. It is easy to see that Φ is well-defined. (In case R is a field, this is already done in Witt (1936), and one can deduce the general case from the field case via functoriality. The case $n = 1$ is treated at length in Nagahara and Nakajima (1971).) We next show surjectivity (this is the main point). Let S/R be a C_{p^n}-Galois extension. By Thm. **0** 6.1, the $R[C_{p^n}]$-module S is finitely generated and projective, hence a direct summand of some free $R[C_{p^n}]$-module. Therefore $\mathrm{H}^1(C_{p^n}, S) = 0$. Now C_{p^n} also acts naturally on $W_m(S)$, and $W_m(S)$ is an extension of $W_{m-1}(S)$ by $W_1(S) = S$, whence (by induction on m, and letting $m = n$) also $\mathrm{H}^1(C_{p^n}, W_n(S))$ vanishes. Since C_{p^n} is cyclic with generator σ, this means:

For all $\underline{x} \in W_n(S)$ with $\mathrm{tr}(\underline{x})=0$, there is $\underline{y} \in W_n(S)$ with $\underline{x} = \sigma(\underline{y}) \dot{-} \underline{y}$.

(Here tr is the norm element $\sum_{i=0}^{p^n-1} \sigma^i \in \mathbb{Z}[C_{p^n}]$; we call it trace now, because the modules under consideration are additive.) We take $\underline{x} = \mathbf{1}$: then $\mathrm{tr}(\underline{x}) = p^n \cdot \mathbf{1} = 0$ because $W_n(S)$ is annihilated by p^n. Therefore one finds $\underline{y} \in W_n(S)$ with $\sigma(\underline{y}) = \underline{y} \dot{+} \mathbf{1}$. Since $\mathbf{1}^P = \mathbf{1}$, and $(.)^P$ is an automorphism, the element $\underline{y}^P \dot{-} \underline{y}$ is stable under σ, hence (direct calculation) $\underline{z} = \underline{y}^P \dot{-} \underline{y}$ is in $W_n(R)$, and we obtain a well-defined σ-equivariant R-algebra map

$$S_{\underline{z}} = R[\underline{Y}]/(\underline{Y}^P \dot{-} \underline{Y} \dot{-} \underline{z}) \longrightarrow S,$$

which maps $\underline{Y}$ to $\underline{y}$, i.e. $\underline{Y}_i$ to $\underline{y}_i$ $(1 \le i \le n)$. By **0** 1.12 this map must be an isomorphism, hence $\Phi(\underline{z})$ is the class of S.

Now we show that Φ is injective. Suppose $S_{\underline{x}}$ is trivial, i.e. isomorphic to the C_{p^n}-fold product of R. Then there is an R-algebra map π: $R[\underline{Y}]/(\underline{Y}^P \dot{-} \underline{Y} \dot{-} \underline{x}) \to R$. We may consider $\pi(\underline{Y})$ as an element of $W_n(R)$, and since π is an R-algebra homomorphisms, we get $\pi(\underline{Y})^P \dot{-} \pi(\underline{Y}) = \underline{x}$.

It remains to see that Φ is a homomorphism. Take $\underline{x}, \underline{y} \in W_n(R)$, and let S be the Harrison product of $\Phi(\underline{x}) = R[\underline{X}]/(\underline{X}^P \dot{-} \underline{X} \dot{-} \underline{x})$ and $\Phi(\underline{y}) = R[\underline{Y}]/(\underline{Y}^P \dot{-} \underline{Y} \dot{-} \underline{y})$. Then S is the fixed subgroup of (σ, σ^{-1}) in $A = \Phi(\underline{x}) \otimes_R \Phi(\underline{y})$. Let the vector $\underline{t} \in W_n(A)$ be defined by $\underline{t} = \bar{\underline{X}} \dot{+} \bar{\underline{Y}}$. Then one sees that all components of $\underline{t}$ are in S, and one obtains a well-defined homomorphism α from $R[\underline{Z}]/(\underline{Z}^P \dot{-} \underline{Z} \dot{-} (\underline{x} \dot{+} \underline{y}))$ to S sending each Z_i to t_i. Then α is σ-equivariant and necessarily already an isomorphism, Q.E.D.

Suppose now that $n \geq 2$, $q = p^n$, $q' = p^{n-1}$. Then $C_{q'}$ is embedded in $C_q = C_{p^n}$ via ι, where ι sends a generator σ' to σ^p. By functoriality of H, there arises a homomorphism ι^*: $\mathrm{H}(R, C_{q'}) \longrightarrow \mathrm{H}(R, C_q)$. On the other hand one has a shift operator s: $W_{n-1}(R) \longrightarrow W_n(R)$ which sends $(x_0, x_1, \ldots, x_{n-1})$ to $(0, x_0, x_1, \ldots, x_{n-2})$. One then has a compatibility between ι^* and s:

Lemma 1.2. *The diagram*

$$\begin{array}{ccc} W_{n-1}(R) & \xrightarrow{\Phi} & \mathrm{H}(R, C_{q'}) \\ s \downarrow & & \downarrow \iota^* \\ W_n(R) & \xrightarrow{\Phi} & \mathrm{H}(R, C_q) \end{array}$$

is commutative.

Proof. Let $x \in W_{n-1}(R)$. Consider the $C_{q'}$-extension $S_{\underline{x}} = R[\underline{Z}]/(\underline{Z}^P \dot{-} \underline{Z} \dot{-} \underline{x})$. From the definition of ι^*, one calculates that

$$\iota^* S_{\underline{x}} \;=\; S_{\underline{x}}^{\{0,1,\ldots,p-1\}} \text{ with } C_q\text{-action } \sigma(s_0, \ldots, s_{p-1}) = (s_1, s_2, \ldots, s_{p-1}, \sigma'(s_0)).$$

Next consider the C_q-extension $S_{s(\underline{x})} = R[\underline{Y}]/(\underline{Y}^P \dot{-} \underline{Y} \dot{-} s(\underline{x}))$. Note that $\underline{x}$ and $\underline{Z}$ are vectors of length $n-1$, and $s(\underline{x})$, $\underline{Y}$ are of length n. As in the preceding proof, it suffices to find a σ-equivariant algebra map f: $S_{s(\underline{x})} \longrightarrow \iota^* S_{\underline{x}}$. We define f via its components $f_0, \ldots, f_{p-1}$: $S_{s(\underline{x})} \longrightarrow S_{\underline{x}}$:

$$f_i((Y_0, \ldots, Y_{n-1})) \;=\; i \cdot (1,0,\ldots,0) \dot{+} (0, \overline{Z}_0, \ldots, \overline{Z}_{n-2}) \in W_n(S_{\underline{x}}). \qquad (0 \leq i \leq p-1)$$

These f_i are actually well-defined, since $(1,0,\ldots 0)^P = (1,0,\ldots,0)$ and $(0, \overline{Z}_0, \overline{Z}_1, \ldots, \overline{Z}_{n-2})^P = s(\overline{Z})^P = s(\overline{Z}^P) = s(\overline{Z} \dot{+} \underline{x}) = (0, \overline{Z}_0, \overline{Z}_1, \ldots, \overline{Z}_{n-2}) \dot{+} s(\underline{x})$. We now calculate as follows: $f_i(\sigma(Y_0, Y_1, \ldots, Y_{n-1})) = f_i((1,0,\ldots,0) \dot{+} (Y_0, Y_1, \ldots, Y_{n-1}) = (1,0,\ldots,0) \dot{+} i \cdot (1,0,\ldots,0) \dot{+} (0, \overline{Z}_0, \overline{Z}_1, \ldots, \overline{Z}_{n-2})$. For $i < p-1$, this equals $f_{i+1}((Y_0, \ldots, Y_{n-1}))$. For $i = p-1$, we use $p \cdot (1,0,\ldots,0) = (0,1,0,\ldots,0)$ (actually $s(\underline{z}^P) = p \cdot \underline{z}$ for all $z \in W_n(R)$) and find the following: $f_i(\sigma(Y_0, Y_1, \ldots, Y_{n-1})) = (0,1,0,\ldots,0) \dot{+} (0, \overline{Z}_0, \overline{Z}_1, \ldots, \overline{Z}_{n-2}) = s((1,0,\ldots,0) \dot{+} (\overline{Z}_0, \overline{Z}_1, \ldots, \overline{Z}_{n-2})) = s(\sigma'(\underline{Z})) = \sigma'(s(\underline{Z})) = \sigma'(f_0((Y_0, \ldots, Y_{n-1})))$. These formulas show that f is σ-equivariant, q.e.d.

Remark 1.3. *The preceding lemma has a counterpart in Kummer theory: if R is p^n-kummerian, and i: $R^*/p^n \longrightarrow \mathrm{H}(R, C_{p^n})$ is the Kummer map (see* **0** *§5), then a similar diagram as in* 1.2 *(replace Φ by i, and s by the p-th power map), is also commutative. This is easy to prove, and has already been used in Chap. V.*

At the end of this §, we review the theory of Witt vectors of infinite length. For any ring A, one can define $W(A)$ as the inverse limit $W_n(A)$. This is again a ring. For us the most important feature is that in many cases $W(A)$ as a ring has an intrinsic characterization.

Definition. a) The ring A is called *perfect* (p-perfect would be more precise), if it has characteristic p (a prime) and the p-power map $A \to A$, $a \longmapsto a^p$, is bijective.

b) Suppose A is perfect with char. p. A ring B with a given isomorphism π: $B/pB \longrightarrow A$ is called a *Witt ring* of A if B is p-adically complete, i.e. $B = \lim_{\leftarrow} B/p^n B$, and the following conditions are satisfied:

(*i*) There is a multiplicative map j: $A \to B$ such that the composite $\pi \cdot \mathrm{can} \cdot j$: $A \to B \to B/pB \to A$ is the identity.

(*ii*) Every $x \in B$ has a unique representation $x = \sum_{\nu=0} p^\nu \cdot j(a_\nu)$, with elements $a_\nu \in A$.

The conditions in this definition are unnecessarily strong (but useful), as shown by the next result:

Theorem 1.4. *Let A be perfect, let B be p-adically complete (see above), and π: $B/pB \to A$ an isomorphism. Then (B,π) is a Witt ring of A iff p is a nonzero-divisor in B. The ring $W(A)$ with π = projection to the zeroth component is a Witt ring of A, and any other Witt ring B of A is isomorphic to $W(A)$. If j is a multiplicative section as in the above definition, then such an isomorphism is afforded by*

$$t\colon W(A) \longrightarrow B,\ (a_0, a_1, \ldots) \longmapsto \textstyle\sum_{\nu=0} p^\nu \cdot j(a_\nu^{p^{-\nu}}) \in B.$$

Moreover, given two Witt rings (B,π) and (B',π') of A, there is exactly one isomorphism t: $B \to B'$ such that t induces $\pi'^{-1}\pi$ modulo p.

Proof. This is much the same as for the classical case where A is a field (see Witt (1936)), and reasonably well-known. The necessity of the condition "p is not a zero-divisor" is easy; we indicate why it is sufficient. Define j: $A \to B$ by $j(a) = \lim_{n\to\infty} (b_n^{p^n})$, where b_n is a preimage under π of $a^{p^{-n}}$. One shows as in the classical case that j is well-defined, and multiplicative, the key fact being the implication "$b \equiv c \pmod p \Rightarrow b^{p^n} \equiv c^{p^n} \pmod{p^{n+1}}$". Condition (*i*) is then immediate. Cf. Mumford (1964) [p. 26.5].

Condition (*ii*) and the remaining statements of the theorem now follow from Mumford (loc.cit.) or Serre (1962) [Thm. II.7 and Prop. II.10].

§2 Patching results

We shall prove here some technical results on patching of Galois extensions which we want to be at hand in later sections. The method we shall use is due to Karoubi in the case of projective modules, and was described by Landsburg (1981). We give simple and direct proofs adapted to our situation.

The framework is as follows: R is a ring in which p is not a zero-divisor, $\hat{R} = \lim_{\leftarrow} R/p^n R$ its p-adic completion, $\iota\colon R \longrightarrow \hat{R}$ the canonical map (which need not be injective), and we define $K = R[p^{-1}]$, $\hat{K} = \hat{R}[p^{-1}]$. *Warning*: K and $\hat{K}$ need not be fields, and $\hat{K}$ is not the p-adic completion of K in the above sense.

The proof of the next result is easy, and omitted:

Lemma 2.1. *Also in* $\hat{R}$, *p does not divide zero, and hence* $\hat{R}$ *embeds into* $\hat{K}$.

Denote the embedding $\hat{R} \longrightarrow \hat{K}$ by ι'.

Lemma 2.2. *The diagram*

$$\begin{array}{ccc} R & \subset & K \\ {\scriptstyle\iota}\downarrow & & \downarrow{\scriptstyle\iota'} \\ \hat{R} & \subset & \hat{K} \end{array}$$

is a fiber product diagram ("cartesian", in another terminology).

Proof. We have to show: Given $y \in \hat{R}$, $k \in K$ with same image in $\hat{K}$, there is exactly one $x \in R$ which goes to y and k. More concretely: k has to be in $R \subset K$, and $\iota(k)$ must be y. Let $y = (y_n) \in \lim_{\leftarrow} R/p^n R$, and choose $N \in \mathbb{N}$ with $p^N \cdot k \in R$. From $\iota'(k) = y$ we get $\iota'(p^N k) = p^N y$, hence $p^N k \equiv p^N y_n \pmod{p^n}$ for all n. Therefore $p^N k$ is divisible by p^N in R (take $n = N$), and this implies $k \in R$ since p is not a zero-divisor in $K \supset R$. The condition $\iota(k) = y$ now follows from $\iota'(k) = y$, q.e.d.

Corollary 2.3. *For every finitely generated projective R-module M, the following is also a fiber product diagram:*

$$\begin{array}{ccccccc} & & M & \subset & K \otimes_R M & = & M[p^{-1}] \\ & & \downarrow & & \downarrow & & \\ \lim_{\leftarrow} M/p^n M & = & \hat{M} & \subset & \hat{K} \otimes_{\hat{R}} \hat{M} & = & \hat{M}[p^{-1}]. \end{array}$$

Proof. By hypothesis, $\hat{M} = \hat{R} \otimes_R M$ and $\hat{K} \otimes_R M = \hat{K} \otimes_{\hat{R}} \hat{M}$. Hence the above diagram is just the diagram in 2.2, tensored with M. Since M is R-flat, it is again cartesian.

Corollary 2.4. *If G is a finite group, and A/R is G–Galois, then we may define $B = \hat{R} \otimes_R A$, $C = K \otimes_R A$, $D = \hat{K} \otimes_R A$; these are G–Galois over $\hat{R}$, K, and $\hat{K}$ respectively, and A is canonically isomorphic to their fiber product:*

$$\begin{array}{ccc} A & \dashrightarrow & C \\ \vdots & & \downarrow \\ B & \longrightarrow & D\,. \end{array}$$

What we want to do now is to reverse this procedure: we want to construct an $A \in \mathrm{H}(R,G)$, if $B \in \mathrm{H}(R,G)$ and $C \in \mathrm{H}(K,G)$ (with some compatibility) are given. We do a slightly more special case first:

Theorem 2.5. *Let C/K be G–Galois (G a finite group), and $\iota_C\colon C \longrightarrow \hat{C} = \hat{K} \otimes_K C$ the canonical map. Suppose moreover that B is an $\hat{R}$–submodule of $\hat{C}$ which is G–Galois over $\hat{R}$ (with the induced structures), and such that $B[p^{-1}] = \hat{C}$. Then the R–module $A = \iota_C^{-1}(B)$ is a G–Galois extension of R with the induced structures, and one gets back B and C as in 2.4: B is canonically isomorphic to $\hat{R} \otimes_R A$, and $C = A[p^{-1}]$.*

Proof. We first show A/R is G–Galois. For this we use the criterion **0** 1.6, i.e. we have to show:

a) $A^G = R$;

b) For all maximal ideals $\mathfrak{M}$ of A, and all $\sigma \in G - \{id\}$, there exists $x \in A$ with $\sigma(x) - x$ not in $\mathfrak{M}$.

a) is not hard: $A^G \subset \iota_C^{-1}(B^G) \cap C^G = \iota_C^{-1}(\hat{R}) \cap K \subset \iota'^{-1}(R)$ with $\iota'\colon K \to \hat{K}$ as in 2.2. By 2.2, $\iota'^{-1}(\hat{R}) = R$. The reverse inclusion $R \subset A^G$ is trivial.

For b) we need two lemmas:

Lemma 2.6. $A[p^{-1}] = C$.

Proof. "$\subset$" is clear. Let $c \in C$; then $\iota_C(c) \in \hat{C} = B[p^{-1}]$, hence there is some $N \in \mathbb{N}$ with $p^N \iota_C(c) \in B$. This implies $\iota_C(p^N c) \in B$, hence $p^N c \in \iota_C^{-1}(B) = A$, i.e. $c \in A[p^{-1}]$.

Lemma 2.7. *$A/p^n A \cong B/p^n B$ canonically for all n.*

Proof. Trivially $p^n A \subset \iota_C^{-1}(p^n B)$; we show the other inclusion. If $y \in C$, $\iota_C(y) \in p^n B$, then also $y/p^n \in C$ and $\iota_C(y/p^n) \in B$. Then by definition, $y/p^n \in A$, which suffices. We now have that the canonical map (induced by ι_C)

$$\varkappa\colon\ A/p^n A = \iota_C^{-1}(B)/\iota_C^{-1}(p^n B) \longrightarrow B/p^n B$$

is injective, and it remains to show that $\varkappa$ is surjective.

Given $b \in B$, we need some $c \in C$ with $b - \iota_C(c) \in p^n B$ (for then $c \in \iota_C^{-1}(B) = A$, and $\varkappa(c + p^n A) = b + p^n B$). In other words, we must show $b \in \iota_C(c) + p^n B$. Now b is in $\hat{C}$ and hence can be written $b = \sum r_i \cdot \iota_C(c_i)$ for some $r_i \in \hat{R}$, $c_i \in C$. By 2.6, all c_i have the form $p^{-N}a_i$, with $a_i \in A$. Moreover one has $r_i = \iota(s_i) + p^{N+n}\rho_i$ for appropriate $s_i \in R$ and $\rho_i \in \hat{R}$ (we denote the map $R \to \hat{R}$ by ι). It follows that

$$b = \sum r_i \cdot \iota_C(c_i) = \sum \iota_C(s_i c_i) + p^{N+n} \sum \rho_i \cdot \iota_C(c_i) = \sum \iota_C(s_i c_i) + p^n \sum \rho_i \iota_C(a_i),$$

and this lies in $\iota_C(C) + p^n B$, q.e.d.

After these preparations, we can check condition b) above for A. Take $\sigma \in G$, $\sigma \neq id$, and $\mathfrak{M} \subset A$ maximal. First case: $p \in \mathfrak{M}$. Then we have $A/\mathfrak{M} \cong (A/pA)/(\mathfrak{M}/pA) \cong (B/pB)/(\mathfrak{M}B/pB)$ by 2.7, and $\mathfrak{M}B/pB$ is maximal in B/pB. Since B/pB is G-Galois over R/pR (reason: B/R is G-Galois by hypothesis, and $R/pR \cong R/pR$), we can use **0** 1.6 the other way and find some $s' \in B/pB$ which is moved by σ mod $(\mathfrak{M}B/pB)$, and then any preimage s of s' in A will do.

Second case: p is not in $\mathfrak{M}$. Then $A/\mathfrak{M} = A[p^{-1}]/\mathfrak{M}A[p^{-1}] = C/\mathfrak{M}C$, and $\mathfrak{M}C$ is a maximal ideal in C. Since by hypothesis C/K is G-Galois, we can find $s' \in C$ which is moved by σ mod $\mathfrak{M}C$. Then any $s \in A$ with s mod $\mathfrak{M}$ a preimage of s' mod $\mathfrak{M}C$ will do.

Hence A/R is indeed G-Galois. We already showed in 2.6 that $C = A[p^{-1}]$, and $B \cong \hat{R} \otimes_R A$ follows from 2.7, since $\hat{R} \otimes_R A \cong \varprojlim A/p^n A$ (A is finitely generated projective over R). Q.E.D.

For applications, it is practical to have a slightly more general version:

Theorem 2.8. *Suppose B' and C are G-Galois extensions of R (of K, respectively), such that $\hat{K} \otimes_{\hat{R}} B'$ and $\hat{K} \otimes_K C$ are isomorphic as G-Galois extensions of K. Then there exists a G-Galois extenson A of R which induces B' (and C) by base change from R to $\hat{R}$ (from R to K, respectively).*

If moreover $G = C_q = \langle\sigma\rangle$ is cyclic and R contains a q-th primitive root ζ of unity (q some power of p), and if B' contains an element x with $\sigma(x) = \zeta \cdot x$ and x^q is the image $\iota(a) \in R^*$ of some $a \in R^*$, then we can choose C and A as in the first paragraph in such a way that A contains an element y with $\sigma(y) = \zeta \cdot y$ and $y^q = a$.

Proof. There is a G-isomorphism φ: $\hat{K} \otimes_{\hat{R}} B' \longrightarrow \hat{K} \otimes_K C = \hat{C}$. Let $B = \varphi(B') \subset \hat{C}$. Then Thm. 2.5 produces a G-extension A of R with $A[p^{-1}] = C$, $\hat{R} \otimes_R A \cong B \cong B'$. Moreover, A is explicitly given as the subset $\iota^{-1}(B)$ of C.

Suppose now we are in the situation described in the second half of the theorem. Then by Kummer theory the C_q-extension $\hat{K} \otimes_{\hat{R}} B' = B'[p^{-1}]$ is isomorphic to the Kummer extension $\hat{K}(q;\iota(a)) = \hat{K}[T]/(T^q - \iota(a))$, with $\sigma(\bar{T}) = \zeta \cdot \bar{T}$, and $x \in B'$ corresponding with $\bar{T}$. (Use **0** 1.12.) Hence we may take C to be the Kummer extension $K[T]/(T^q - a)$, and $\varphi(x) = 1 \otimes T$ affords a G-isomorphism

$K \otimes_R B' \longrightarrow K \otimes_K C$. Then $y = \overline{T} \in C$ is in $\iota^{-1}(\varphi(B'))$, since $\iota(y)$ is just the element $1 \otimes \overline{T} = \varphi(x) \in \hat{K}[T]/(T^q - \iota(a))$. Then clearly $y \in A$, $y^q = a$, and $\sigma(y) = \zeta \cdot y$.

§ 3 Kummer theory without the condition "$p^{-1} \in R$"

In this section we fix a prime p ($p = 2$ is allowed), a power p^n of p, and we consider a ring R which contains a root ζ_n of the p^n-th cyclotomic polynomial, and in which p is not a zero-divisor. We shall call ζ_n a primitive p^n-th root of unity. Write ζ for ζ_n, and let λ ($= \lambda_n$ if necessary) be defined to be $1 - \zeta$. It is well-known that the $(p-1)p^{n-1}$-th power of λ is associated to p in the ring R, since this already holds in the ring $\mathbb{Z}[\zeta]$.

We can define higher unit groups in R: Suppose p is not a unit in R. For $a = (p-1)^{-1} \cdot p^{-n+1} \cdot b \in ((p-1)p^{n-1})^{-1}\mathbb{N}$, i.e. $b \in \mathbb{N}$, we define

$$\mathrm{U}_a(R) \;=\; \{u \in R^* \,|\, 1 - u \text{ is divisible by } \lambda^b\}.$$

One sees easily that these sets are subgroups of R^*. For example, a unit $u \in R^*$ belongs to $\mathrm{U}_m(R)$ ($m \in \mathbb{N}$), iff $1 - u$ is divisible by p^m. (We shall really need fractional subscripts a.)

The main theme are again Galois extensions of R with group C_{p^n}, where $C_{p^n} = \langle\sigma\rangle$ is cyclic of order p^n. We will introduce soon a certain subgroup $\mathrm{F}_n(R)$ of $\mathrm{H}(R, C_{p^n})$. It contains $\mathrm{NB}(R, C_{p^n})$, and is more amenable to calculations in the present setting. (The disadvantage is that the definition of $\mathrm{F}_n(R)$ is slightly less natural.) For semilocal rings R, all three groups $\mathrm{NB}(R, C_{p^n})$, $\mathrm{F}_n(R)$, and $\mathrm{H}(R, C_{p^n})$ coincide.

For the definition of $\mathrm{F}_n(R)$ we use a generalization of the map π: $\mathrm{H}(R, C_{p^n}) \longrightarrow \mathrm{Pic}(R)[p^n]$ from usual Kummer theory. In the next definition and proposition, we essentially follow Haggenmüller (1985).

Definition and Proposition 3.1. *Define for every character* χ: $C_{p^n} \to \langle\zeta\rangle$ *a map* $\det_\chi$: $\mathrm{H}(R, C_{p^n}) \longrightarrow \mathrm{Pic}(R)$ *by*

$$\det_\chi(A) = \textit{class of the } R\textit{-module } A^\chi = \{x \in A \,|\, \sigma(x) = \chi(\sigma) \cdot x\}.$$

Then $\det_\chi$ *is a well-defined group homomorphism (in particular,* A^χ *is an invertible* R*-module). This was proved in* Borevič (1979) *and can also be shown by faithfully flat descent. If* $p^{-1} \in R$ *and* $\chi(\sigma) = \zeta$*, then* $\det_\chi$ *is just the map* π *of* **0** §5.

Furthermore, one has canonical isomorphisms $A^\chi \otimes_R A^\psi \cong A^{\chi\psi}$ for all characters χ and ψ (copy the proof in **0** *§5). Hence A^χ is free for* ***all*** *characters χ if and only A^{χ_1} is free for some primitive character χ_1: $C_{p^n} \to \langle\zeta\rangle$.*

This proposition motivates the following definition:

Definition. The group $F_n(R) \subset H(R, C_{p^n})$ of *determinantally free* (*d-free* for short) C_{p^n}-extensions of R is defined to be the kernel of $\det_\chi$, where χ is the character of C_{p^n} given by $\chi(\sigma) = \zeta$.

Remark. If $A \in NB(R, C_{p^n})$, then $A \cong R[C_{p^n}]$ as $R[C_{p^n}]$-modules, hence $A^\chi \cong R[C_{p^n}]^\chi$, and it is quite easy to check that the latter R-module is free of rank 1. Therefore $NB(R, C_{p^n})$ is a subgroup of $F_n(R)$.

The next step is the construction of a canonical homomorphism φ_n from $F_n(R)$ to R^*/p^n (recall this is our shorthand for R^*/R^{*p^n}). The idea is that φ_n should be a partial inverse of the map i of Kummer theory, which is no longer defined.

Definition. Let $A \in F_n(R)$. Then A^χ is free cyclic, generated by ξ, say. Define $\varphi_n(A)$ (or, more correctly, $\varphi_n([A])$!) to be the class of ξ^{p^n} in R^*/p^n.

Lemma 3.2. *φ_n is a well-defined group homomorphism, and functorial in R.*

Proof. Let $q = p^n$. The power ξ^q is in R, since $\sigma(\xi^q) = (\chi(\sigma)\cdot\xi)^q = \zeta^q\cdot\xi^q = \xi^q$. By the above proposition, the multiplication map $(A^\chi)^{\otimes q} \to R$ is an isomorphism (note that for the trivial character ε, $A^\varepsilon = R$). Hence ξ^q is invertible in R. If one changes ξ by a factor from R^*, then the power ξ^q will only change by a factor in R^{*q}, so φ_n is well-defined. It is also a group homomorphism: if A' is also in $F_n(R)$, and ξ' an R-generator of A'^χ, then (as one recalls) the Harrison product B of A and A' is the fixed subring of $\mathrm{Ker}(\mu: C_{p^n} \times C_{p^n} \to C_{p^n})$ in $A \otimes_R A'$, and one checks that $\xi \otimes \xi'$ is an R-generator of B^χ. From this the formula $\varphi_n(A)\cdot\varphi_n(A') = \varphi_n(B)$ follows easily. The functoriality of φ_n is equally straightforward, q.e.d.

The underlying idea is that φ_n should be injective, so that knowledge of $\mathrm{Im}(\varphi_n)$ gives complete information about $F_n(R)$, our object of interest. While this is not true in complete generality, it works in a case which often occurs, as described in the next lemma. It seems remotely possible but probably very cumbersome to find a description of $\mathrm{Ker}(\varphi_n)$ in general. For the case $n = 1$, see Haggenmüller (1985). We will not pursue this any further.

Lemma 3.3. *If R is integrally closed in $R[p^{-1}]$, then φ_n is injective.*

Proof. Write $\varphi_{R,n}$ instead of φ_n for a moment. Over $R[p^{-1}]$ we can use Kummer theory, hence $F_n(R[p^{-1}])$ is just the kernel of $\pi_{R[p^{-1}]}$ and therefore equal to the image of $i = i_{R[p^{-1}]}$. One also sees that $\varphi_{R[p^{-1}],n}$ is just the inverse of i, hence injective. Consider now the commutative diagram

$$\begin{array}{ccc} F_n(R) & \xrightarrow{\varphi_{R,n}} & R^*/p^n \\ j \downarrow & & \downarrow \\ F_n(R[p^{-1}]) & \xrightarrow{\varphi_{R[p^{-1}],n}} & R[p^{-1}]^*/p^n , \end{array}$$

where j is the base extension map. By Harrison (1965) [Thm. 5], j is injective, cf. Cor. **0** 4.2. Therefore $\varphi_{R,n}$ is also injective. As a **corollary** to the proof we get, looking at Harrison (loc.cit.) or **0** 4.2 again: If $\varphi_n(A) = \overline{u} \in R^*/p^n$, then A is the integral closure of $B = R[X]/(X^q - u)$ in $B[p^{-1}]$. Q.E.D.

Example. Let $p^n = 2$, $R = \mathbb{Z}_2$. One can then show $\mathrm{Im}(\varphi_1)$ consists of all $u \in \mathbb{Z}_2^*$ with $u \in U_2(\mathbb{Z}_2)$, i.e. $u \equiv 1 \pmod 4$. Cf. the examples at the end of §5.

There is another way of interpreting the map φ_n:

Definition and Remark 3.4. *Let* $E_n(R)$ *be the preimage of* $\mathrm{Im}(\varphi_n) = \varphi_n(F_n(R))$ *under the canonical surjection* $R^* \longrightarrow R^*/p^n$. *Then an element* $x \in R^*$ *is in* $E_n(R)$ *iff there exists* $A \in F_n(R)$ *and* $\xi \in A$ *such that* $\sigma(\xi) = \zeta\cdot\xi$ *and* $\xi^q = x$ *(recall* $q = p^n$*).*

Proof: "If" is clear by definition, and "only if" can be seen as follows: if $x \in E_n(R)$, then we find $A \in F_n(R)$ and $\xi' \in A$ such that $\sigma(\xi) = \zeta\cdot\xi$ and $y = \xi'^q$ equals $u^q x$ for some $u \in R^*$. It now suffices to set $\xi = u^{-1}\xi'$.

We add another explanatory remark: For any $x \in R^*$ one may look at the R-algebra $A_x = R[T]/(T^q - x)$. The group $C_q = C_{p^n}$ operates on A_x by $\sigma(\overline{T}) = \zeta\cdot\overline{T}$, but one can show that A_x is *not* C_{p^n}-Galois over R unless $p^{-1} \in R$. To obtain a C_{p^n}-extension, one has in general to enlarge A_x slightly (in the example above, this amounts to the adjunction of $(1-\overline{T})/2$, as can be shown). This enlarging process is not always possible. We claim that $x \in R^*$ is in $E_n(R)$ iff there is a C_{p^n}-invariant embedding of A_x in a C_{p^n}-Galois extension A of R. Suppose first that $x \in E_n(R)$, so there is A/R C_{p^n}-Galois, $\xi \in A^\chi$, $\xi^q = x$ (Remark 3.4). Then $\overline{T} \longmapsto \xi$ affords the required embedding. If, conversely, A_x is embedded in $A \in H(R, C_{p^n})$, then A^χ contains $\overline{T}$, and $\overline{T}$ is already a unit in A_x. It follows easily (e.g. by localizing) that then A^χ is already R-generated by $\overline{T}$. From the definition of φ_n, we obtain that $x\cdot R^{*q} = \varphi_n(A)$, hence $x \in E_n(R)$.

To conclude this section, we establish a result which provides an a priori (lower) estimate of the group $E_n(R)$. The proof uses §2 of this chapter for the passage from the complete to the general case.

Theorem 3.5. *As always in this* §, *suppose* p *is a non-zero-divisor in* R, *and* R *contains the primitive* p^n*-th root* ζ *of unity. We consider the higher unit group*

$$\mathrm{U}_{n,+}(R) \overset{def}{=} \mathrm{U}_{n+\frac{1}{p-1}}(R) = \{x \in R^* \,|\, 1-x \text{ divisible by } p^n \cdot \lambda^{p^{n-1}} \text{ in } R\}$$

(recall $\lambda = 1-\zeta$, p is associated to $\lambda^{(p-1)p^{n-1}}$). Then $\mathrm{U}_{n,+}(R)$ is contained in $\mathrm{E}_n(R)$.

Proof. We formulate two assertions. The second will result easily from §2, and the two assertions together easily imply the theorem. The central matter is, then, the proof of the first assertion.

Assertion 1: *Thm. 3.5 holds if R is p-adically complete, i.e. $R = \varprojlim R/p^n R$.*

Assertion 2: *Let $\hat{R}$ be the p-adic completion of R, ι: $R^* \longrightarrow \hat{R}^*$ the canonical map. Then $\iota^{-1}(\mathrm{E}_n(\hat{R})) = \mathrm{E}_n(R)$.*

Proof of **3.5** modulo the two assertions: Let $x \in \mathrm{U}_{n,+}(R)$. Then certainly $\iota(x) \in \mathrm{U}_{n,+}(R)$, hence by the first assertion, $\iota(x) \in \mathrm{E}_n(R)$. The second assertion gives $x \in \mathrm{E}_n(R)$, q.e.d.

Proof of Assertion **2**: The inclusion "$\supset$" follows from functoriality. The other inclusion follows from (the second half of) Thm. 2.8 and Remark 3.4.

Finally, we turn to the *proof* of the first assertion. Suppose henceforth that *R* is *p*-adically complete. We begin with a lemma which is in principle well-known.

Lemma 3.6. *Every $x \in \mathrm{U}_{n+1,+}(R)$ has a p^n-th root y in $\mathrm{U}_{1,+}(R)$.*

Proof. Set $\lambda_p = 1-\zeta_p$ with ζ_p the p^{n-1}-th power of ζ. Then λ_p is associated to $\lambda^{p^{n-1}}$ in *R* (this holds already in $\mathbb{Z}[\zeta]$). We have $x = 1+d$ with *d* divisible by $p^{n+1}\cdot\lambda_p$. We use the binomial series:

$$(1+d)^{p^{-n}} = \sum_{i=0}^{\infty} \binom{p^{-n}}{i} \cdot d^i \quad .$$

To justify the use of this series, we have to show that all its terms make sense in *R*, that it converges, and that its limit is congruent to 1 (mod $p\lambda_p$). Let $q = p^n$, and let v_p be the usual *p*-adic valuation of integers. We then have;

$$\begin{aligned} v_p\binom{p^{-n}}{i} &= -v_p(i!) + v_p(q^{-1}) + v_p(q^{-1}-1) + \ldots + v_p(q^{-1}-i+1) \\ &\geq (1-i)/(p-1) - i\cdot n \qquad \text{for } i \geq 1. \end{aligned}$$

Since d^i is divisible by $(p^{n+1}\lambda_p)^i$, we obtain that $\binom{p^{-n}}{i}\cdot d^i$ is at least divisible by $p^i\cdot\lambda_p$ for $i \geq 1$. Hence all terms of the series make sense in *R*, the series converges, and its limit is congruent to its zeroth term (which is 1) modulo $p^1\lambda_p$, q.e.d.

Remark. One may also give a slightly longer proof using an iterative procedure for finding the root *y*.

Now we prove Assertion 1 by induction over n. For $n = 1$, this is basically a result of Childs (1977), cf. also Haggenmüller (1985) and Waterhouse (1987): Let $x \in U_{1,+}(R)$. Then the algebra $A_x = R[T]/(T^p - x)$ $(\sigma(\overline{T}) = \zeta\cdot\overline{T})$ can be enlarged to a C_p-extension A of R by Childs (1977) [Thm. 2.5]. To be precise, one has to take $A = R\left[(1-\overline{T})/\lambda_p\right] \subset A_x[p^{-1}]$.

Now suppose $n \geq 2$, and let $r = p^{n-1}$. By Lemma 3.6, x has an r-th root y in R, $y \equiv 1 \pmod{p\lambda_p}$. Let $S = R[X]/(X^q - x)$, $\sigma(\overline{X}) = \zeta\cdot\overline{X}$. We have to construct a C_{p^n}-extension S' into which S embeds σ-equivariantly. One proceeds in two steps. First, fix a generator τ of C_p, and let $\zeta_p = \zeta^r$ (a primitive p-th root of unity). Let $T_1 = R[Y]/(Y^p - y)$ with C_p-action $\tau(\overline{Y}) = \zeta_p\cdot\overline{Y}$. Let T be the product of r copies of T_1, and define an embedding $f\colon S \longrightarrow T$ by

$$f(X) = (\overline{Y}, \zeta\overline{Y}, \ldots, \zeta^{r-1}\overline{Y}).$$

The map f is well-defined since the p^n-th power of the right hand side equals $(Y^p)^r = x$, and f becomes an isomorphism on adjoining p^{-1}, by the Chinese Remainder Theorem. Therefore f is an embedding. If we let σ act on T by

$$\sigma(t_0, t_1, \ldots, t_{r-1}) = (t_1, t_2, \ldots, t_{r-1}, \tau(t_0)), \qquad (*)$$

then f is σ-equivariant. By the case $n = 1$, we may extend T_1 to a C_p-Galois extension T_2. Let S' be the product of r copies of T_2. Of course, f is also an embedding $S \to S'$. The definition (*) equally makes sense for S'. One can see now that S' is exactly ι^*T_2, where ι is the inclusion $C_p \to C_{p^n}$, $\tau \longmapsto \sigma^r$ (see **0** §3). Hence S'/R is a C_{p^n}-Galois extension, q.e.d.

§ 4 The main result, and Artin–Hasse exponentials

Let us keep the notation of §3. In particular, R contains the primitive p^n-th root ζ of unity, and $p \in R$ is not a zerodivisor. In §3 we defined a homomorphism φ_n from the group $F_n(R)$ of d-free C_{p^n}-extensions to the group R^*/p^n. The preimage of $\mathrm{Im}(\varphi_n)$ in R^* was denoted $E_n(R)$. Let now ψ be the canonical surjection

$$\psi\colon R^* \longrightarrow R^*/U_{n,+}(R) \qquad \text{(see §3 for the definition of } U_{n,+}).$$

Let us denote $\psi(E_n(R))$ by $D_n(R)$. Then Thm. 1.4 tells us that $D_n(R) = E_n(R)/U_{n,+}(R)$, and it seems reasonable to deal with D_n instead of E_n. In this context, one should recall that $D_n(R)$ completely determines $F_n(R)$ whenever R is integrally closed in $R[p^{-1}]$, by Lemma 3.3.

The first step is again a reduction to the complete case. Let $\hat{R}$ be the p-adic completion of R. Then $\iota\colon R \longrightarrow \hat{R}$ induces a canonical map

$$\iota_n:\ R^*/U_{n,+}(R) \longrightarrow \hat{R}^*/U_{n,+}(\hat{R}).$$

A moment's thought will convince the reader that ι_n is monic (although ι need not be monic). Since every element of R which is a unit mod p is itself already a unit (geometric series, or the like), the group $\hat{R}^*/U_{n,+}(\hat{R})$ is canonically isomorphic with the unit group of $\hat{R}/p^n\lambda_p\hat{R} \cong R/p^n\lambda_p R$. One sees already that the range of ι_n is the unit group of a ring which is often a finite ring (e.g. if R is a ring of integers in a number field). The second Assertion in the proof of Thm. 3.5 gives immediately:

Lemma 4.1. $D_n(R) = \iota_n^{-1}(D_n(R))$.

Motivated by this lemma, we shall assume for the whole rest of the chapter that R is *p-adically complete*. The corresponding results for R noncomplete are obtained from those for R complete by appealing to 4.1. Note that (as just explained) the group $D_n(R)$ is essentially a subgroup of $(R/p^n\lambda_p)^*$.

We now state the first main result of this chapter. Fix a family of primitive p^i-th roots of unity in $\mathbb{C}$ with $\zeta_{i+1}{}^p = \zeta_i$ for all i. If $i \le n$ and ζ_i is to be considered as an element of R, then $\zeta_i = \zeta^{p^{n-i}}$ (ζ was a fixed primitive p^n-th root of 1 in R).

Theorem 4.2. *There exist explicitly given power series $f_i \in \mathbb{Z}_p[\zeta_i][[X]]$ $(1 \le i < \infty)$, independent of R, such that the ν-th coefficient of f_i converges p-adically to zero with $\nu \to \infty$ for all i (which implies that $f_i(r)$ is defined for all $r \in R$) and such that*

$$D_n(R) = \{f_n(r_n)\cdot f_{n-1}(r_{n-1})^p \cdot \ldots \cdot f_1(r_1)^{p^{n-1}} \cdot r_0^{p^n} \mid r_0, \ldots, r_n \in R\}$$

as a subgroup of $R^*/U_{n,+}(R)$. Explicitly one has for $n \ge 1$:

$$f_n(X) = \exp(g_n(X)),$$

$$\begin{aligned} g_n(X) &= -\ p^n\eta X \\ &\quad -\ (p^n\eta + p^{n-1}\eta^p)\,X^p \\ &\quad -\ (p^n\eta + p^{n-1}\eta^p + p^{n-2}\eta^{p^2})\,X^{p^2} \\ &\quad -\ \ldots \\ &= -\sum_{\nu=0}^{\infty}\Big(\sum_{\mu=0}^{\nu} p^{n-\mu}\eta^{p^\mu}\Big)X^{p^\nu}, \end{aligned}$$

where $\eta = \eta_n$ is a special element of $\mathbb{Z}_p[\zeta_n]$ which will be defined later.

Remarks. a) From the explicit series for $g_n(X)$, it is *not* evident that the ν-th coefficient goes p-adically to zero, but we shall see it in the course of the proof.

b) It is clear from the statement of the theorem that it is sufficient to know f_i modulo $p^i\lambda_p$ (because then one will know $f_i{}^{p^{n-i}}$ modulo $p^n\lambda_p$). In particular, one may always replace f_l by some polynomial over $\mathbb{Z}_p[\zeta_l]$ in calculations, and even by a polynomial with coefficients in $\mathbb{Z}[\zeta_l]$.

c) We shall see below that η_1 is associated in R to $\lambda_p = 1-\zeta_1$ for $n = 1$. Using this, one sees easily that $g_1(X) \equiv 0 \pmod{p\lambda_p}$ and $f_1(X) \equiv 1 \pmod{p\lambda_p}$, which means that $f_1(r_1)^{p^{n-1}}$ is congruent to 1 mod $p^n\lambda_p$. Hence the f_1 term is actually superfluous in the above description of $\mathrm{D}_n(R)$, which is practical to know for calculations. In the theory, we shall retain the f_1 term for formal reasons.

Theorem 4.2 has its roots in the paper of Hasse (1936). Our result is both more general and more explicit, but Hasse's formalism developed in loc. cit. is quite essential for our proofs.

We shall first do some reduction steps for the proof of 4.2, and then introduce the mentioned formalism, to wit: Artin-Hasse exponentials. (The need for such a formalism will be a bit clearer by then.) The main part of the proof, and examples, will follow in §5.

Lemma 4.3. *(Recall R is supposed to be p-adically complete.)* $\mathrm{H}(R, C_{p^n}) = \mathrm{F}_n(R)$, *i.e. all C_{p^n}-extensions of R are determinantally free.*

Proof. We even show that every $A \in \mathrm{H}(R, C_{p^n})$ has a normal basis (cf. Remark following 2.1). By Nakayama's Lemma, it suffices to show that $A' = A/pA$ has a normal basis modulo $R' = R/pR$. Now R' has characteristic p, and therefore all elements $1-\sigma$, $\sigma \in C_{p^n}$, are nilpotent in the group ring $R'[C_{p^n}]$, i.e. the augmentation ideal I of $R'[C_{p^n}]$ is nilpotent. Hence, again by Nakayama, $x \in A'$ is an $R'[C_{p^n}]$-generator of A' iff it is an R'-generator of A'/IA'. In the proof of 1.1 we saw that IA' is exactly the kernel of the trace tr: $A' \to R'$, and tr is surjective by **0** 1.10. Thus, x will be an R'-generator of A'/IA' iff $\mathrm{tr}(x)$ is a generator of R'. Such an x certainly exists, again by **0** 1.10. To be explicit: any $x \in A'$ with $\mathrm{tr}(x) = 1$ generates a normal basis of A', q.e.d.

The starting point of the argument is now the fact that the canonical map

$$\alpha\colon \mathrm{H}(R, C_{p^n}) \longrightarrow \mathrm{H}(R', C_{p^n}) \qquad \text{with } R' = R/pR$$

is an isomorphism because R is p-complete. For a proof of this result, see Greither and Haggenmüller (1982), or EGA IV, 18.1.2. From Lemma 4.3 and Thm. 1.1 we obtain the following diagram for n > 1:

$$\begin{array}{ccccccc}
W_{n-1}(R') & \xrightarrow[\sim]{\Phi} & \mathrm{H}(R', C_{p^{n-1}}) & \xrightarrow[\sim]{\alpha^{-1}} & \mathrm{H}(R, C_{p^{n-1}}) & \xrightarrow{\varphi_{n-1}} & R^*/\mathrm{U}_{n-1,+}(R)\cdot R^{*p^{n-1}} \\
{\scriptstyle s}\downarrow & & {\scriptstyle \iota^*}\downarrow & & {\scriptstyle \iota^*}\downarrow & & {\scriptstyle p\text{-th power}}\downarrow \\
W_n(R') & \xrightarrow[\sim]{\Phi} & \mathrm{H}(R', C_{p^n}) & \xrightarrow[\sim]{\alpha^{-1}} & \mathrm{H}(R, C_{p^n}) & \xrightarrow{\varphi_n} & R^*/\mathrm{U}_{n,+}(R)\cdot R^{*p^n}
\end{array} .$$

The rightmost vertical arrow is well-defined, since $U_{n-1,+}(R)^p \subset U_{n,+}(R)^p$. The diagram commutes by 1.2 and 1.3.

We can now see two things: First, to determine the image of φ_n (i.e. to determine $D_n(R)$) amounts to calculating explicitly the map $\varphi_n\alpha^{-1}\Phi$ which we shall denote inv_n. Second, the diagram helps us to accomplish this by induction over n. More precisely: Consider the subset $V_n = \{(r,0,\ldots,0) \mid r \in R'\}$ of $W_n(R)$. (This is not a subgroup!) Then we get $W_n(R') = V_n + s(W_{n-1}(R'))$, whence $\mathrm{Im}(\mathrm{inv}_n) = \mathrm{inv}_n(V_n) + \mathrm{Im}(\mathrm{inv}_{n-1})^p$ by the diagram. By induction over n we get:

Proposition 4.4. $\mathrm{Im}(\mathrm{inv}_n) \;=\; \mathrm{inv}_n(V_n)\cdot\mathrm{inv}_{n-1}(V_{n-1})\cdot\ldots\cdot\mathrm{inv}_1(V_1)^{p^{n-1}}$.

On the other hand, by definition $\mathrm{Im}(\mathrm{inv}_n) = \mathrm{Im}(\varphi_n) = D_n(R)/\overline{R^{*p^n}}$ (the overbar means reduction mod $U_{n,+}(R)$). Therefore Theorem 4.2 is a consequence of the preceding proposition, and the next theorem, whose proof is deferred to §5:

Theorem 4.5. *For $n \geq 1$, $\mathrm{Im}(\mathrm{inv}_n) = \{\overline{f_n(r)} \mid r \in R\}$ $\left(\subset R^*/R^{*p^n}\cdot U_{n,+}(R)\right)$, and $\overline{f_n(r)}$ only depends on r modulo p.*

The power series f_n has been defined in 4.2, modulo the quantity η_n. See §5.

Moreover, for $n = 1$, inv_1 is the trivial homomorphism.

The rough idea of proof is that $\mathrm{inv}_n(r,0,\ldots,0)$ is a sort of exponential function of r. (The Witt ring is an additive group, and inv_n goes to a multiplicative group.) To make this vague idea more precise, one has to use so-called Artin-Hasse exponentials. One wants to give a meaning to the formal expression $(1+x)^y$, where x and y are elements of some p-adically complete ring R, and a power of x is divisible by p (i.e. x^m goes to zero for $m \to \infty$). The naive approach "$(1+x)^y = \exp(y\cdot\log(1+x))$" does not work. In Hasse (1936) the required exponential is introduced for the case that R is the Witt ring of a perfect field. This is too special for our purposes. In the following preparations, R can be any p-adically complete ring R in which p does not divide zero. Later on, we work with Witt rings of perfect rings of char. p, and their cognates.

The following material is straight from Hasse's paper. Let R be p-complete, p not a zero-divisor in R, and let $\mathfrak{M} = \{x \in R \mid \text{some power of } x \text{ is in } pR\}$. Define $K = R[p^{-1}] \supset R$. For all $x \in \mathfrak{M}$ we can form the series over K

$$(4.6) \qquad L(1-x) \;=\; \sum_{\tau=0}^{\infty} \frac{x^{p^\tau}}{p^\tau},$$

which converges in K, i.e. per def. it lies and converges in $p^{-N}R$ for some large $N \in \mathbb{N}$. For all $x \in \mathfrak{M}$ we also have the logarithm series which converges in K:

$$(4.7) \qquad -\log(1-x) \;=\; \sum_{n=1}^{\infty} \frac{x^n}{n}.$$

Therefore

$$(4.8) \qquad -\log(1-x) = \sum_{(p,m)=1} \frac{1}{m} L(1-x^m).$$

Let μ: $\mathbb{N} \longrightarrow \{-1,0,1\}$ be the Moebius function. As in loc. cit. we define

$$(4.9) \qquad P(1-x) = \prod_{(p,m)=1} (1-x^m)^{\frac{\mu(m)}{m}}.$$

This series makes sense (one develops the m^{-1}-th power by the binomial series, which causes no problems as long as m is prime to p), and $P(1-x) \in 1+\mathfrak{M}$. Using that the sum $\sum_{d|m} \mu(d)$ is 1 for $m = 1$ and zero for $m > 1$, we get

$$(4.10) \qquad L(1-x) = -\log P(1-x).$$

In 4.6 – 4.10, one may also replace x by the indeterminate X and thus obtain valid equations involving $P(1-X) \in R[[X]]$ and $L(1-X)$, $\log(1-X) \in K[[X]]$. By looking at the constant and the linear term one finds that $P(1-X) \equiv 1-X \pmod{X^2}$. Hence there are power series $\psi(X)$, $\eta(X) \in R[[X]]$, congruent to $X \pmod{X^2}$, such that

$$(4.11) \qquad P(1-X) = 1-\psi(X), \text{ and } P(1-\eta(X)) = 1-X.$$

By 4.10 we get

$$(4.12) \qquad -\log(1-\psi(X)) = L(1-X), \quad -\log(1-X) = L(1-\eta(X)),$$

and the analogous formulas if we substitute $x \in \mathfrak{M}$ back for X.

Remark. One may think of $P(1-x)$ as a modified exponential: the series $L(1-X)$ is, as it were, the p-power part of the log series, and 4.10 suggests that P is somehow "the non-p-power part of exp". In particular, P converges much better than exp. By the same token, the series η is a modified logarithm (second formula in 4.12).

For the definition of the exponential we need additional conditions on R. We repeat from §1:

Definition. A ring A is *perfect* (= *p-perfect*), if char$(A) = p$, and the p-th power map $x \longmapsto x^p$ is bijective on A.

In §1 we showed that every perfect ring A has a Witt ring B, which is unique up to unique isomorphims, and can be taken to be the ring $W(A)$ of (infinite) Witt vectors over A. There is a multiplicative section j: $A \to B$ of the surjection π: $B \to B/pB = A$. If we take $B = W(A)$, then $j(a) = (a, 0, \ldots)$.

We now define the *Artin–Hasse exponential* for exponents in the Witt ring B of an arbitrary perfect ring A:

Definition. a) Let $a = \sum_{\nu=0} j(a_\nu)\cdot p^\nu \in B$. The power series $(1-X)^a$ is given by

$$(1-X)^a \;=\; \prod_{\nu=0}^{\infty} P(1-j(a_\nu)\cdot\eta(X))^{p^\nu} \;\in\; B[[X]].$$

The product converges, since the ν-th factor is congruent to 1 mod $(p,X)^\nu$, and the constant term of $(1-X)^a$ is 1.

b) Let R be any p-complete overring of B, $x \in \mathfrak{M} \subset R$. Then we define

$$(1-x)^a \;=\; (1-X)^a\big|_{X=x} \in R.$$

The principal properties of this exponential are summarized in the following:

Theorem 4.13. (Hasse) *Let A and B be as above.*

(0) *For $r \in \mathbb{Z}_p \subset B$, the series $(1-X)^r$ just defined agrees with $(1-X)^r$ defined via the binomial expansion.*

(1) $(1-X)^a \;\equiv\; 1-aX \quad (mod\ X^2)$ *for $a \in B$.*

(2) $(1-X)^a(1-X)^b \;=\; (1-X)^{a+b}$ *for $a,b \in B$.*

(3) $(1-X)^{ar} = ((1-X)^a)^r$ *for $a \in B$ and $r \in \mathbb{Z}_p$.*

(**Warning**. *The formula $(1-X)^{ar} = ((1-X)^r)^a$ is far from true.*)

Proof. In the number-theorical case, this has been done by Hasse (1936). There is practically no change in our situation; for the convenience of the reader we give the argument.

(1) is seen directly as follows. We have mod X^2: $(1-X)^a \equiv \prod_\nu (1-j(a_\nu)X)^{p^\nu} \equiv 1-\sum_\nu j(a_\nu)p^\nu\cdot X = 1-aX$.

For (2) and (3) we show that the logarithms of both sides are equal. (It is easily seen that $\log\colon 1+X\cdot B[[X]] \longrightarrow B[p^{-1}][[X]]$ is injective.) The ring B carries an automorphism P determined by $j(a)^P = j(a^P)$ and $p^P = p$. (To see this, identify B with the ring $W(A)$ of Witt vectors, see beginning of §1). The summation of series in $B[p^{-1}][[X]]$ in the following calculation will be no problem, since the ν-th term will be zero mod X^{p^ν}. We calculate:

$$\begin{aligned}
-\log((1-X)^a) \;&=\; -\sum_{\nu=0}^{\infty} p^\nu \log P(1-j(a_\nu)\cdot\eta(X)) \\
&=\; \sum_{\nu=0}^{\infty} p^\nu L(1-j(a_\nu)\cdot\eta(X)) \\
&=\; \sum_{\nu=0}^{\infty} p^\nu\cdot \sum_{\rho=0}^{\infty} p^{-\rho} j(a_\nu)^{P^\rho}\eta(X)^{p^\rho} \\
&=\; \sum_{\rho=0}^{\infty}\Big(\sum_{\nu=0}^{\infty} p^\nu\cdot j(a_\nu^{P^\rho})\Big)\cdot p^{-\rho}\eta(X)^{p^\rho}
\end{aligned}$$

$$= \sum_{\rho=0}^{\infty} a^{P^\rho} p^{-\rho} \eta(X)^{p^\rho}.$$

From the last equation (2) and (3) follow, since $a^P + b^P = (a+b)^P$, and $(ar)^P = a^P r^P = a^P r$ (reason for the last equality: P is the identity on $\mathbb{Z}_p = W(\mathbb{F}_p)$, since it induces the Frobenius of $\mathbb{F}_p$, which is identity). For $a = r \in \mathbb{Z}_p$, the last expression above simplifies to $r \cdot L(1-\eta(X)) = r \cdot \log(1-X) = \log((1-X)^r)$, where the r-th power is the usual one, given by the binomial series. This gives (0), q.e.d.

Corollary 4.14. *The rules* (0) - (3) *hold also if we substitute for* X *any* $x \in \mathfrak{M} \subset R$, *where* R *is any* p*-complete overring of* B, *and* $\mathfrak{M} = \{y \in R \mid \exists t \in \mathbb{N}: y^t \in pR\}$.

(NB. The exponent a *must always be in the Witt ring* B, *but* R *might be* $B[\zeta_n]$ *for example.)*

Proof: immediate consequence of 4.13.

In the next section we shall use this formalism in order to calculate the image of the Witt vector $(\bar{r}, 0, \ldots, 0) \in W_n(R/pR)$ under the map inv_n: $W_n(R/pR) \longrightarrow R^*/\mathrm{U}_{n,+}(R) \cdot R^{*p^n}$, for every p-adically complete ring R in which p is not a zero-divisor. Unfortunately, R' need not be perfect, which means that the formalism of Artin-Hasse exponentials is not yet available over R. Therefore we must consider several auxiliary rings, which makes matters complicated.

§ 5 Proofs and examples

Let as before R be a ring which contains the primitive p^n-th root $\zeta = \zeta_n$ of unity, and in which p is not a zero-divisor. Let $R' = R/pR$. Our objective is the proof of Thm. 4.5 (and, concomitantly, the definition of the quantities η_n).

In §4, preceding the statement of 4.5, we defined a map inv_n which associates to every Witt vector x of length n over R' a class in R^* modulo $\mathrm{U}_{n,+}(R) \cdot R^{*p^n}$. This map is given by the following recipe: x gives by Artin-Schreier theory a C_{p^n}-extension S'/R'. Lift this to a C_{p^n}-extension S/R (this is the hard part!), and finally find a unit $\xi \in S^*$ with $\sigma(\xi) = \zeta \cdot \xi$. Then inv_n = class of the element ξ^{p^n} (that element must indeed lie in R^*). The explicit description of this whole process is now given by the following:

Theorem 4.5. (final version): *Define* $\eta = \eta_n \in \mathbb{Z}_p[\zeta_n]$ *by*

$$\eta_n = \eta(1-\zeta_n), \text{ where } \eta(X) \text{ is as in §4.}$$

(This makes sense since a power of $\lambda = 1-\zeta_n$ *is divisible by p.)*
Then for $x = (\bar{r},0,\ldots,0) \in W_n(R')$:

$$\mathrm{inv}_n(x) = \overline{f_n(r)} \in R^*/\mathrm{U}_{n,+}(R)\cdot R^{*p^n}.$$

Moreover, inv_1 *is the trivial homomorphism.*

The idea of proof is, in a nutshell: Take an "Artin–Schreier element" y for S', i.e. $y \in W_n(S')$ with $\sigma(y) = y \dot{+} 1$. Show that S is a Witt ring of S' and may be thus identified with $W(S')$. Let ξ be the Artin–Hasse exponential $(1-\lambda)^{y_1}$, with y_1 a lift of y to $S \cong W(S')$. Modulo a certain power of λ, one obtains $\sigma(\xi) = (1-\lambda)^{\sigma(y_1)} \equiv (1-\lambda)^{y_1+1} = \zeta\cdot\xi$ (recall $\lambda = 1-\zeta$ by definition). By a suitable choice of the lift y_1 one achieves equality in the place of congruence, i.e. ξ is now a Kummer element for S over R, and it "only" remains to calculate z, the p^n-th power of ξ. One problem is that S' need not be perfect, i.e. $(1-\lambda)^{y_1}$ is undefined. Another problem is to get a manageable expression for the result, and it is here that we go further than in Hasse's theory.

We have to work with a certain relatively small base ring R_0. Let us first see how we obtain a simplification in doing so.

Lemma 5.1. *Let* R_0 *be the p-adic completion of* $\mathbb{Z}[\zeta_n, X]$ *(*ζ_n *a prim.* p^n*-th root of 1 in* $\mathbb{C}$*). The map* inv_n *is defined also for* R_0*, and in order to prove 4.5, it suffices to show that* $\mathrm{inv}_n((\bar{X},0,\ldots,0))$ *is the class of* $f_n(X)$ *in* $R_0^*/\mathrm{U}_{n,+}(R_0)\cdot R_0^{*p^n}$.

Proof. With the notation of 4.5 (final version), we consider the continuous ring homomorphism β_r: $R_0 \longrightarrow R$ uniquely determined by $\beta_r(X) = r$. Clearly inv_n is functorial in R (since Φ and α are, see §4). Therefore $\mathrm{inv}_n((r,0,\ldots,0)) = \beta_r(\mathrm{inv}_n((\bar{X},0,\ldots,0))$. If we know that $\mathrm{inv}_n((\bar{X},0,\ldots,0) = \overline{f_n(X)}$, then we are done since $\beta_r(f_n(X)) = f_n(r)$, q.e.d.

Using this lemma, and going through the definition of inv_n, we see that we are reduced to proving the following: ($R_0' = R_0/pR_0$)

Principal Claim *If we let* $R_n = \alpha^{-1}\Phi_{R_0'}((X,0,\ldots,0))$ *(in words:* R_n *is the lift of the* C_{p^n}*-extension of* R_0' *which belongs by Thm. 1.1 to the Witt vector* $(X,0,\ldots,0)$*), then there exists an element* $z \in R_n^*$ *such that:*

$$z \equiv 1 \pmod{\lambda}, \quad \sigma(z) = \zeta\cdot z \text{ (i.e. z is a Kummer element), and:}$$

$$z^{p^n} \equiv f_n(X) \pmod{p^n\lambda_p.}$$

(Recall $\lambda = 1-\zeta$*, and* $\lambda_p = 1-\zeta_1$, $\zeta_1 = \zeta^{p^{n-1}}$.*)* *Moreover* $f_1(X) \equiv 1 \bmod p\lambda_p$.

In proving this claim, it turns out that the ring R_0 is too small. It is of the form $R_0 = Z[\zeta_n]$ with $Z = \mathbb{Z}[X]^\wedge$, and $Z' = Z/pZ = \mathbb{F}_p[X]$ is not perfect (X has no p-th root, for instance), so Z cannot be a Witt ring. This practically forces us to replace Z by the ring

$$B = \mathbb{Z}_p[X^{p^{-\infty}}]^\wedge = \mathbb{Z}_p[X, X^{1/p}, X^{1/p^2}, \ldots]^\wedge,$$

where $^\wedge$ denotes p-adic completion. It is not difficult to see that $B' = B/pB = \mathbb{F}_p[X^{p^{-\infty}}]$ is perfect. Therefore B is a Witt ring of B', and $j(X) = X$ because $X^{p^{-n}} \in B$ goes to $X^{p^{-n}} \in B'$ for all n, see 1.4 . Let B_n' be the C_{p^n}-Galois extension of B' which belongs to $(X,0,\ldots,0)$ by 1.1 (Artin-Schreier), and let B_n be a pre-image of B_n' under α, i.e. B_n is a C_{p^n}-extension of B which induces B_n'/B'.

Lemma 5.2. *B_n' is again perfect.*

Proof. For any B'-module M, let ${}_pM$ denote the abelian group M with B' operating via Frobenius: $b*m = b^p m$. One then has ${}_pM = B' \otimes_p M$, where $B' \otimes_p -$ means base change along the inverse of Frobenius $F_{B'}^{-1}: B' \to B'$. Then for any B'-algebra M, the Frobenius F_M on M defines a B'-algebra map $M \to {}_pM$. Hence F_M is a homomorphism from one C_{p^n}-extension of B' to another, hence an isomorphism by **0** 1.12, q.e.d.

By Lemma 5.2, B_n is a Witt ring of $B_n' = B_n/pB_n$ (it is easy that p is again a nonzero-divisor in B_n). Let D and D_n denote $B[\zeta]$ and $B_n[\zeta]$, respectively. The adjunction of ζ has to be understood formally: adjoin a root of the p^n-th cyclotomic polynomial. Then D_n is C_{p^n}-Galois over D. We now state a theorem and a proposition, which together imply the Principal Claim. The main work will then be involved in the proof of the theorem.

Theorem 5.3. *There exists an element $z \in D_n^*$, $z \equiv 1 \pmod{\lambda}$, $\sigma(z) = \zeta\cdot z$, such that $z^{p^n} = f_n(X) \in R_0 \subset D$.*

Proposition 5.4. *If $v \in D^*$, and $v^p \in R_0 \subset D$, then already $v \in R_0$.*

Proof of the **implication "5.3 & 5.4 ⇒ Principal Claim"**. By Lemma 4.3, R_n is d-free over R_0, i.e. there exists $\xi \in R_n^*$ with $\sigma(\xi) = \zeta\cdot\xi$, and hence $\xi^{p^n} \in R_0^*$. Take z as in Thm. 5.3, and consider the quotient $v = \xi/z$. Since $\sigma(v) = v$, we have $v \in D^*$. Moreover, $v^{p^n} = \xi^{p^n}/f_n(X) \in R_0^*$. By Prop. 5.4 (applied n times), we even have $v \in R_0^*$. Therefore $z \in R_n^*$, and the principal claim follows from 5.3. (The concluding statement of the principal claim concerning f_1 will be a byproduct in the proof of 5.3.)

Proof of **Prop. 5.4.** Recall $D = \mathbb{Z}_p[X^{p^{-\infty}}, \zeta]^\wedge$ with $^\wedge$ standing for p-adic completion. (It does not matter whether we first adjoin ζ and then complete, or vice versa.)

Let $v \in D^*$. Mod p^2, we have $D/p^2D = \bigcup_{l \in \mathbb{N}} (\mathbb{Z}/p^2)[X^{p^{-l}}, \zeta]$, hence v is congruent mod p^2 to some v_0 in $Z = \mathbb{Z}_p[X^{p^{-e}}, \zeta]^\wedge$ for an appropriately large e. This v_0 is also a unit, and we may write $v = v_0(1+p^2u)$ for some $u \in D$. Since by hypothesis $v^p \in R_0 = \mathbb{Z}_p[X,\zeta]^\wedge$, we find that $y = (1+p^2u)^p$ is in Z. Hence $\log y = p \cdot \log(1+p^2u)$ is in Z since Z is p-adically complete, and $Z \cap pD = pZ$. By that argument again, $\log(1+p^2u)$ is in Z, and taking exp again (which we may do), we get $1+p^2u \in Z$, hence $v \in Z$, and automatically then $v \in Z^*$. Now $Z = R_0[X^{p^{-e}}]$, and we want to show for $v \in Z^*$: $v^p \in R_0$ forces $v \in R_0$.

One sees easily that it suffices to treat $e = 1$. Then Z becomes a Kummer extension of degree p upon inverting p. As for the field case, all elements of $Z[p^{-1}]$ whose p-th power lies in $R_0[p^{-1}]$ are of the form $\sqrt[p]{X}^s \cdot w$, with $0 \le s < p$ and $w \in R_0[p^{-1}]$. Looking at the p-denominators in such an equation $v = \sqrt[p]{X}^s \cdot w$, we see that w must be already in R_0. Now $s \neq 0$ is impossible because v is a unit, hence $v = w \in R_0$, q.e.d.

Proof of **Thm. 5.3.** We first prove two lemmas.

Lemma 5.5. *If S is any ring, $x \in S$, then:*

$$(1-x)^p \equiv 1 \pmod{x^p} \text{ if } x^p \text{ divides } px;$$

$$(1-x)^p \equiv 1 \pmod{px} \text{ if } px \text{ divides } x^p.$$

Proof: Easy, writing out the binomial expansion of $(1-x)^p$.

Lemma 5.6. *Let $n \le N \in \mathbb{N}$, and T' a perfect ring with Witt ring T. If $\Delta = \sum_{\nu \ge 0} j(a_\nu) \cdot p^\nu$ ($a_\nu \in T'$) is an element of T, and $\Delta \equiv 0 \pmod{p^n}$ (which just means that $a_0, \ldots, a_{N-1}$ vanish), then*

$$(1-\lambda)^\Delta \equiv 1 \pmod{p^{N-n+1}} \text{ in } T[\zeta].$$

(N.B. $(1-\lambda)^\Delta$ is an Artin–Hasse exponential.)

Proof. In the product formula (§4) which defines $(1-\lambda)^\Delta$, the factors with $\nu = 0, \ldots, \nu = N-1$ are equal to 1 and may be ignored. Now for any $b \in T$ one has $1 - b \cdot \eta(\lambda) \equiv 1 \pmod{\lambda}$, hence also $P(1 - b \cdot \eta(\lambda)) \equiv 1 \pmod{\lambda}$ by 4.11, and therefore $P(1-b\cdot\eta(\lambda))^{p^n} \equiv 1 \pmod{p\lambda_p}$ by Lemma 5.5 (recall that $\lambda^{p^{n-1}}$ is associated to λ_p and λ_p^{p-1} is associated to p). Using 5.5 again, we find $P(1-b\cdot\eta(\lambda))^{p^N} \equiv 1 \pmod{p^{N-n+1}\lambda_p}$ for each $N \ge n$. This gives the desired congruence for all factors of the product defining $(1-\lambda)^\Delta$, q.e.d.

Back to the proof of 5.3. This is somewhat complicated. It is our aim to find elements $z_N \in D_n$ for all $N \ge n$, such that:

(z_N) converges in D_n (even $z_N \equiv z_{N+1} \pmod{p^{N-n+1}}\ \forall N$);

$$z_N \equiv 1 \pmod{\lambda};$$
$$\sigma(z_N) \equiv \zeta \cdot z_N \pmod{p^{N-n+1}}; \text{ and}$$
$$z_N^{p^n} \equiv f_n(X) \pmod{p^{N-n+1}}.$$

It is then clear that $z = \lim(z_N)$ satisfies all conditions of Theorem 5.3. In the process we shall need C_{p^N}-extensions for all $N \geq n$. (This phenomenon occurs already in Hasse's paper.)

Let σ_N be a fixed generator of C_{p^N} ($N \in \mathbb{N}$). The groups C_{p^N} form a projective system via $\sigma_{N+1} \longmapsto \sigma_N$. Let it be agreed that letters with double underbar denote Witt vectors of length N, ordinary underbar meaning length n. For $N \geq n$, we let B_N' be the C_{p^N}-extension

$$B_N' = B'[\underline{\underline{\Theta}}]/(\underline{\underline{\Theta}}^p \dot{-} \underline{\underline{\Theta}} \dot{-} \underbrace{(X,0,\ldots,0)}_{(N-1 \text{ zeros})}), \qquad \sigma_N(\Theta) = \Theta \dot{+} \mathbf{1}.$$

Then one has natural embeddings $B_n' \subset B_{n+1}' \subset \ldots$, which are compatible with the Galois action (one may say that $\bigcup B_N'$ is a $\mathbb{Z}_p$-extension of B_n'). One can lift this whole situation to the ground ring B by Greither and Haggenmüller (1982) [2.1, p.244]: There are C_{p^N}-extensions B_N of B, $B_n \subset B_{n+1} \subset \ldots$, which induce the B_N' when one goes modulo p again. By Lemma 5.2, the B_N' are all again perfect, hence B_N is a Witt ring for B_N' for all $N \geq n$. Therefore by 1.4 there are isomorphisms $\gamma_N \colon B_N \longrightarrow W(B_N')$, uniquely determined by the condition that they induce the identity on B_N'. By this uniqueness, one also knows that γ_N is C_{p^N}-equivariant.

Let $\vartheta_N \in B_N$ such that the first N components of $\gamma_N(\vartheta_N)$ give just $\overline{\underline{\underline{\Theta}}} \in W_N(B_N')$. Then we have

$$(*) \qquad \sigma_N(\vartheta_N) \equiv \vartheta_N + \mathbf{1} \pmod{p^N},$$

because $\gamma_N(\sigma_N(\vartheta_N) - \vartheta_N - 1) \in W(B_N')$ goes to zero if we project it to $W_N(B_N')$ and hence lies in $p^N W(B_N')$. Now we define $D_N = B_N[\zeta]$. By 4.14, the Artin-Hasse exponential with base $(1-x)$, $x \in D_N$, and exponent $a \in B_N$ is defined in case a power of x is divisible by p. Hence we may define

$$(**) \qquad y_N = \zeta^{\vartheta_N} = (1-\lambda)^{\vartheta_N} \in D_N.$$

It is clear from the definitions that the exponential commutes with σ_N (since everything is define via convergent power series, and the map j, see 1.4). From (*) and 4.13 (2) we get that

$$\sigma_N(y_N) = \zeta^{\sigma_N(\vartheta_N)} = \zeta^1 . \zeta^{\vartheta_N} . \zeta^{\Delta},$$

with $\Delta = \sigma_N(\vartheta_N) - \vartheta_N - 1 \in p^N B_N$. By Lemma 5.6 we now obtain

$$(***) \qquad \sigma_N(y_N) \equiv \zeta \cdot y_N \pmod{p^{N-n+1}}.$$

In particular, the automorphism $\tau = \sigma_N^{p^{N-n}}$ fixes the class of y_N in $D_N/p^{N-n+1}D_N$. Consider now the Galois extensions D_N over D_n, $D_N/p^{N-n+1}D_N$ over $D_n/p^{N-n+1}D_n$, with group $G = \langle\tau\rangle$. It follows that y_N is fixed under G, and hence y_N is in the image of D_n in $D_N/p^{N-n+1}D_N$, which means that there exists $z_N \in D_n$, $z_N \equiv y_N$ (mod $p^{N-n+1}D_N$). We consequently obtain (since $\sigma = \sigma_n$ agrees with σ_N on D_n):

$$\sigma(z_N) \equiv \zeta\cdot z_N \pmod{p^{N-n+1}D_n}.$$

From the construction one sees that $\vartheta_{N+1} \equiv \vartheta_N$ (mod p^N), hence by Lemma 5.6: $y_{N+1} \equiv y_N$, and also $z_{N+1} \equiv z_N$ (mod $p^{N-n+1}D_n$). The congruence $z_N \equiv 1$ (mod λ) is a direct consequence of 4.13 (1). Hence all that is left to show is

$$z^{p^n} = f_n(X) \quad \text{with } z = \lim_{N\to\infty} z_N.$$

To prove this, note first: Since $\sigma(z) = \zeta\cdot z$, the element $b = z^{p^n}$ is certainly in B by Galois theory. It suffices now to show for all $N \geq n$: $y_N^{p^n} \equiv f_n(X)$ (mod $p^{N-n+1}D$).

Since $y_N \equiv 1$ (mod λ) by construction, we have $y_N^{p^n} \equiv 1$ (mod $p\lambda_p$) by Lemma 5.5. We shall now show that also $f_n(X) \equiv 1$ (mod $p\lambda_p$). (NB. This will also give the statement that inv_1 is the trivial map, see the final version of Thm. 4.5 at the beginning of this §.) Since it will be needed, we repeat a formula from the proof of 4.13: For $a = \sum_\nu j(a_\nu)p^\nu \in B$, $x \in D$, some power of x divisible by p, we have

$$(*) \qquad \log(1-x)^a = \Sigma_{\rho\geq 0}\, a^{P^\rho}\frac{\eta(x)^{p^\rho}}{p^\rho},$$

where P is the lift of the Frobenius automorphism of $B' = B/pB$ to B, and j the multiplicative section of $B \to B'$. We defined η to be $\eta(\lambda)$. Letting $x = \lambda$, and $a = 1$, we obtain

$$(**) \qquad 0 = \log(\zeta) = \log(1-\lambda)^1 = \Sigma_{\rho\geq 0}\frac{\eta^{p^\rho}}{p^\rho}.$$

(The fact $\log\zeta = 0$, which is well-known for p-adic fields, follows from $\log 1 = 0$ and the fact that $D[p^{-1}]$ has no p-torsion.) To save space, define $e_\rho = \eta^{p^\rho}/p^\rho$ for $\rho \geq 0$. Then (**) says that $\Sigma_{\rho\geq 0}e_\rho = 0$. Note that the convergence of that series is easy to see a priori. Recall now the definition of $g_n(X)$ from Thm. 4.2 and rewrite it, using (**):

$$g_n(X) = -\sum_{k=0}^{\infty} p^n\Big(\sum_{\rho=0}^{k} e_\rho\Big)\cdot X^{p^k} = \sum_{k=0}^{\infty} p^n\Big(\sum_{\rho=k+1}^{\infty} e_\rho\Big)\cdot X^{p^k}.$$

Now we also know that the coefficients of $g_n(X)$ converge p-adically to zero.

One sees easily that all quantities $p^n e_\rho$ are in $\mathbb{Z}_p[\zeta]$ and therefore have a well-defined p-adic value $v_p(p^n e_\rho)$ (we take the normed valuation, i.e. $v_p(p) = 1$). I-n more detail, the lowest value $v_p(p^n e_\rho)$ is attained (simultaneously!) for $\rho = n-1$ and $\rho = n$: one has $v_p(p^n e_{n-1}) = v_p(p\cdot\eta^{p^{n-1}}) = v_p(p\lambda_p)$, and also $v_p(p^n e_n) = v_p(\eta^{p^n}) = v_p(p\lambda_p) = p/(p-1)$. (**Remark**. It is not a coincidence that the minimum value is attained twice: otherwise the sum of all e_ρ could never vanish!) In particular, $g_n(X)$

is congruent to 0 mod $p\lambda_p$. Hence $f_n(X) = \exp(g_n(X))$ is well-defined, i.e. the exponential converges and there are no denominators involved. Moreover $f_n(X)$ is congruent to 1 mod $p\lambda_p$.

After these preparations we can show $y_N^{p^n} \equiv f_n(X)$ (mod p^{N-n+1}) Since both $y_N^{p^n}$ and $f_n(X)$ are $\equiv 1$ (mod $p\lambda_p$), it suffices to show $\log(y_N^{p^n}) \equiv \log f_n(X)$ (mod p^{N-n+1}), since exp∘log is the identity on the group of units congruent to 1 mod $p\lambda_p$ (a weaker congruence would be enough). Since $\log(y_N^{p^n}) = p^n\cdot\log(y_N) = p^n\cdot\log((1-\lambda)^{\vartheta_N})$, formula (*) gives

$$\text{(!)} \qquad \log(y_N^{p^n}) \equiv p^n\cdot\sum_{\rho=0}^{\infty}\vartheta_N^{p^\rho}\frac{\eta^{p^\rho}}{p^\rho} \equiv p^n\cdot\sum_{\rho=0}^{\infty}\vartheta_N^{p^\rho}e_\rho \pmod{p^{N-n+1}}.$$

Now $\vartheta_N^{\,P} \equiv \vartheta_N + X \pmod{p^N}$ because $\gamma_N(\vartheta_N^{\,P}) = \Theta^P = \Theta + (X,0,\dots,0) = \Theta + \gamma_N(X)$. Abusing notation, we consider here the isomorphism γ_N: $B_N \to W(B_N')$ as an epimorphism $B_N \to W_N(B_N')$. The equality $\gamma_N(X) = (X,0,\dots,0)$ holds because $X \in B$ is the multiplicative representative for $X \in B' = \mathbb{F}_p[X^{p^{-\infty}}]$, in short: $j(X) = X$.

By iteration we get $\vartheta_N^{\,P^\rho} \equiv \vartheta_N + X + X^p + \dots X^{p^\rho}$ (use that P, the lift of Frobenius, actually maps X to X^p). If we insert this in formula (!),we obtain

$$\log(z^{p^n}) \equiv p^n\Big(\sum_{\rho=0}^{\infty}e_\rho\Big)\cdot\vartheta_N + p^n\cdot\sum_{k=0}^{\infty}\Big(\sum_{\rho=k+1}^{\infty}e_\rho\Big)\cdot X^{p^k} \pmod{p^{N-n+1}}.$$

The first sum on the right is zero by (**), and the second is precisely $g_n(X)$. This concludes the proof of Thm. 5.3. Thus, Thm. 4.5 is also proved.

Corollary 5.7. *The element z constructed in the proof of* 5.3 *has the following additional property: The group* $\Gamma = \{\gamma_a \mid a \in (\mathbb{Z}/p^n)^*\}$, $\gamma_a(\zeta) = \zeta^a$, *operates on D by B-algebra automorphisms, and on* D_N *by* B_N*-algebra automorphisms. Then for each* $a \in \mathbb{Z}-p\mathbb{Z}$, *the element* $\gamma_a(z)/z^a$ *is fixed under the action of* C_{p^n} *(and hence in D).*

Proof. It suffices to show that for each $N \geq n$ the element $w_N = \gamma_a(y_N)/y_N^a$ is fixed under C_{p^N} modulo p^{N-n+1}. To prove this, we need two simple observations: (a) the action of C_{p^N} on D_N commutes with the Γ-action since D_N/D is induced by base change from B_N/B; (b) Artin–Hasse exponentials commute with the action of Γ. We now may calculate as follows:

$$\begin{aligned}
\sigma_N(w_N) &= \sigma_N\gamma_a(y_N)/\sigma_N(y_N)^a \\
&= \gamma_a\sigma_N(\zeta^{\vartheta_N})\big/\sigma_N(\zeta^{\vartheta_N a}) && \text{(by 4.13 (3))} \\
&= \gamma_a(\zeta^{\sigma_N(\vartheta_N)})\big/\zeta^{\sigma_N(\vartheta_N)\cdot a} && (\,\sigma_N(a) = a\,) \\
&= \gamma_a(\zeta^{\vartheta_N+1})\big/\zeta^{\vartheta_N a+a} && \text{(since } \sigma_N(\vartheta_N) = \vartheta_N+1) \\
&= \big(\gamma_a(\zeta^{\vartheta_N})/\zeta^{\vartheta_N a}\big)\cdot\gamma_a(\zeta)/\zeta^a \\
&= w_N\cdot\zeta^a/\zeta^a \quad = w_N, \quad \text{Q.E.D.}
\end{aligned}$$

The rest of this section is devoted to **examples**. Let always R be p-adically complete, p not a zero-divisor in R, and $\zeta \in R$ a primitive p^n-th root of unity.

First case: $n = 1$.

Here Thm. 4.2 says that $\mathrm{D}_1(R) = \{\overline{f_1(r)\cdot u^p} \mid r \in R,\ u \in R^*\}$, $\overline{}$ denoting reduction modulo $\mathrm{U}_{1,+}(R)$. But it was remarked after 4.2, and proved in the proof of 5.3, that $f_1(X) \equiv 1 \bmod p\lambda_p$, so we may forget f_1. Therefore the subgroup $\mathrm{E}_1(R) \subset R^*$ is as small as it could possibly be in the light of Thm. 3.5, to wit:

$$\mathrm{E}_1(R) = \mathrm{U}_{1,+}(R)\cdot R^{*p}, \text{ hence}$$
$$\mathrm{Im}(\varphi_1) = \text{image of } \mathrm{U}_{1,+}(R) \text{ in } R^*/R^{*p}.$$

This is essentially the result of Childs (1977) (since R is complete, $\mathrm{H}(R,C_p) = \mathrm{F}_1(R) = \mathrm{NB}(R,C_p)$). Note that we used one half of that result in the proof of Thm. 3.5, which means that we have a new proof only for the other half.

Second case: $n = 2$. (For a different treatment of C_{p^2}-extensions of rings R in which p does not divide zero, see Kersten (1983).)

Here it is worthwile to have a better look at the coefficients $s_\nu = -p^2(e_0+\ldots+e_\nu)$ of X^{p^ν} in $g_2(X)$. Let $u \sim v$ mean that u and are associated in $\mathbb{Z}_p[\zeta]$ (ζ is now a primitive p^2-th root of 1). Then one has:

$$\begin{aligned} p^2e_0 &= p^2\eta, \\ p^2e_1 &= p^2\eta^p/p = p\cdot\eta^p \sim p\cdot\lambda_p, \\ p^2e_2 &= p^2\eta^{p^2}/p^2 = \eta^{p^2} \sim p\cdot\lambda_p, \\ p^2e_3 &= p^2\eta^{p^3}/p^3 = p^{-1}\cdot\eta^{p^3} \sim p^p\cdot\lambda_p. \end{aligned}$$

Since $s_\nu = p^2(e_{\nu+1}+e_{\nu+2}+\ldots)$, this tells us that $p^2\lambda_p$ divides s_ν for $\nu \geq 2$. Thus we have the following approximation to $g_2(X)$:

$$\begin{aligned} g_2(X) &\equiv -p^2e_0X - p^2(e_0+e_1) \\ &\equiv -p^2\eta X - (p^2\eta+p\eta^p)X^p \quad (\mathrm{mod}\ p^2\lambda_p). \end{aligned}$$

One sees (by separating the cases $p = 2$, $p \neq 2$) that in the series $\exp(g_2(X))$ all terms except 1 and $g_2(X)$ are zero modulo $p^2\lambda_p$, and we also get an approximation

$$(*) \quad f_2(X) \equiv 1 - p^2\eta X - (p^2\eta+p\eta^p)X^p \quad (\mathrm{mod}\ p^2\lambda_p).$$

Let us now first consider the case $p = 2$. We know $\eta \equiv \lambda \pmod{\lambda^2}$. Now $\lambda = 1-i$ ($i^2 = -1$), $\lambda_p = 1-(-1) = 2$, and $\lambda^2 = -2i \sim 2$. Moreover $\eta^2 \equiv \lambda^2 \pmod 4$, so we may safely replace all occurences of η in the formulo for $f_2(X)$ $(\mathrm{mod}\ p^2\lambda_p)$ by λ. After a short calculation we obtain

$$f_2(X) \equiv 1 - 4(1-i)X - 4X^2 \quad (\mathrm{mod}\ p^2\lambda_p \sim 8).$$

This implies (note that we may forget the factor f_1 in the statement of 4.2!):

$$D_2R) = \{ \overline{(1+4(1-i)r+4r^2)\cdot u^4} \mid r \in R,\ u \in R^*\} \subset R^*/U_{2,+}(R),$$

or equivalently

$$\begin{aligned} E_2(R) &= \{ (1+4(1-i)r+4r^2)\cdot u^4 \mid r \in R,\ u \in R^*\}\cdot U_{2,+}(R) \\ &= \{ (1+4(1-i)r+4r^2)\cdot u^4 \mid r \in R,\ u \in R^*\}\cdot(1+8R). \end{aligned}$$

Let us turn back to a general prime p (still $n = 2$). A similar argument as above shows that it suffices to know η modulo λ_p (in order to know $f_2(X)$ modulo $p^2\lambda_p$ which is all we want). Define the *truncated log series* by

$$\log_t(1-X) = -\sum_{\nu=1}^{p-1} \frac{X^\nu}{\nu} \in \mathbb{Z}_p[X].$$

Lemma 5.8. *η is congruent to* $-\log_t(1-\lambda)$ (mod λ_p).

Proof. We know that $\eta(X)$ is a power series in X with coefficients in R, and

$$-\log(1-X) = L(1-\eta(X)) = \eta(X) + \eta(X)^p/p + \eta(X)^{p^2}/p^2 + \ldots,$$

and therefore

$$-\log_t(1-X) = \eta(X) + H(X),$$

with H a power series of order p in $R[p^{-1}][[X]]$. Hence H has coefficients in R. When we substitute λ for X, we get $-\log_t(1-\lambda) = \eta + H(\lambda)$, and $H(\lambda)$ is zero (mod $\lambda^p \sim \lambda_p$), q.e.d.

One may now substitute $-\log_t(1-\lambda)$ for η in formula (*). This gives a rather explicit description of $f_2(X)$, and of $D_2(R)$ for every p-adically complete ring R.

Concluding Remark. For $n \geq 3$ the calculations seem to become very involved. It is conceivable, however, that further simplifications in the formula for f_n are possible.

§ 6 Application: Generic Galois extensions

In this final section, we sketch a synthesis of the results of §1-5 of this chapter, and the descent techniques presented earlier in these notes. In particular, we construct a kind of generic C_{p^n}-extension such that the ground ring need neither contain p^{-1} nor ζ_n. (Cf. II §4.) This material is mostly taken from Greither (1989b). Fix a prime p, and a natural number $n \geq 1$.

Let us briefly recall some objects introduced in the last two sections: to begin with, the rings $R_0 = \mathbb{Z}_p[X,\zeta_n]^\wedge$ (ζ_n a prim. p^n-th root of 1), $B = \mathbb{Z}_p[X^{p^{-\infty}}]^\wedge$ (which is a Witt ring of the perfect ring $\mathbb{F}_p[X^{p^{-\infty}}]$), and $D = B[\zeta_n]$. Thus, R_0 is a subring of B.

For each $N \geq n$, we have also C_{p^N}-extensions $B_N \supset B$ (and $D_N = B_N[\zeta_n] \supset D$). The extension B_N/B is characterized via Artin-Schreier theory by the property:

B_N' ($= B_N/pB_N$) is the C_{p^N}-extension of $B' = B/pB$ given by

$$B_N' \;=\; B'[\underline{\underline{\Theta}}]/(\underline{\underline{\Theta}}^p \dot{-} \underline{\underline{\Theta}} \dot{-} (X,0,\dots,0)) \quad \text{(notation of §1 and §5).}$$

We proved in (5.3): There exists a unit $z \in D_n$ which is a Kummer element for D_n/D (i.e. $\sigma(z) = \zeta_n z$), and with the property $z^{p^n} = f_n(X) \in R_0$.

The base ring D is too large in two respects. First, it contains ζ_n; second, its factor ring mod p is not even of finite type over $\mathbb{F}_p$. It is intuitively clear that $Z = \mathbb{Z}_p[X]^\wedge$ would be a better base ring, with a view towards generic extensions. Descending the extension D_n/D from D to B is no problem, because we already know that B_n is a solution. We want to descend to an extension Z_n/Z, however, and this descent from D to Z can be understood better and more explicitly, if we go via the intermediate ring $Z[\zeta_n]$, instead of going via B. Write ζ for ζ_n.

Proposition 6.1. *a) There exists a C_{p^n}-extension Z_n/Z with $D_n = D \otimes_Z Z_n$.*

b) The "intermediate" C_{p^n}-extension $Z_n[\zeta]/Z[\zeta]$ is determinantally free, and φ_n of it is the class of z^{p^n} in $R_0^*/R_0^{*p^n}$.

Proof. a) D_n/D comes by base change from B_n/B. Thus it suffices to show that the latter comes form a C_{p^n}-extension Z_n/Z. Now B_n'/B is an Artin-Schreier extension (explicitly given just above), and it is evident that it descends to a C_{p^n}-extension Z_n' of $Z' = Z/pZ$: it suffices to define

$$Z_n' \;=\; Z'[\underline{\Theta}]/(\underline{\Theta}^p \dot{-} \underline{\Theta} \dot{-} (X,0,\dots,0)) \quad (\text{note } X \in Z' = \mathbb{F}_p[X]).$$

Let Z_n be the (unique) lift of Z_n' to a C_{p^n}-Galois extension of Z. (The existence and uniqueness of such a lifting was already discussed and used in §4.) Then, again by the uniqueness of liftings mod p, Z_n gives B_n upon base extension from Z to B. Note that this argument gives no information at all how to *construct* Z_n.

b) By the remarks preceding the proposition, $\varphi_{n,D}(D_n) = [z^{p^n}] \in D^*/D^{*p^n}$. On the other hand, $Z_n[\zeta]$ must be determinantally free over $Z[\zeta]$ by Lemma 4.3. It hence suffices to see that the canonical map

$$Z[\zeta]^*/p^n \longrightarrow D^*/p^n \qquad (\text{"}\ldots/p^n\text{" is short for "mod } p^n\text{-th powers"})$$

is injective. In other words we need: A unit u of $Z[\zeta]$ which equals a p^n-th power v^{p^n}, $v \in D^*$, is already a p^n-th power in $Z[\zeta]$. Note that $Z[\zeta] = R_0$. Applying Lemma 5.4 n times, we get $v \in Z[\zeta]$, which gives what we want.

Corollary 6.2. *The extension $Z_n[\zeta]$ of $Z[\zeta] = R_0$ can be obtained as follows: Let $b = z^{p^n} \in R_0^*$, form the extension $A = R_0[Y]/(Y^{p^n} - b)$, and take the integral closure of A in $A[p^{-1}]$.*

Proof. It is an easy exercise to show that R_0 is integrally closed in $R_0[p^{-1}]$. Hence the corollary follows from the corollary to the proof of Lemma 3.3.

We now intend to show two things. First, the extension Z_n/Z is in some sense generic; and second, it can be described quite explicitly by descent theory.

For the first task, we of course have to elaborate what we mean by "generic". Let **C** be the category of all p-adically complete (commutative) rings, with continuous ring homomorphisms as morphisms. It is fairly easy to see that Z, the p-adic completion of $Z[X]$, is free on the element X in the category **C**.

Definition. Let $Z' \in$ **C**.

a) A C_{p^n}-extension Z_n/Z' is called *generic*, if for all $S \in$ **C** and all $T \in H(S, C_{p^n})$, there exists a **C**-morphism $\varphi = \varphi_{T/S}$: $Z' \longrightarrow S$ with $T \cong S \otimes_\varphi Z_n$, i.e. T/S is obtained from Z_n/Z' by base extension along φ: $Z' \to S$.

b) A C_{p^n}-extension Z_n/Z' is called *quasi-generic*, if for any $S \in$ **C** and all $T \in H(S, C_{p^n})$, there exists a **C**-morphism $\varphi = \varphi_{T/S}$: $Z' \longrightarrow S$ such that T is isomorphic to the Harrison product $(S \otimes_\varphi Z_n)$ times $\iota^*(B)$ for some $B \in H(C_{p^{n-1}}, S)$, where ι: $C_{p^{n-1}} \to C_{p^n}$ is the inclusion. In other words: T is obtainable by base extension from Z_n/Z' up to some extension which is induced from a $C_{p^{n-1}}$- extension.

(In applications, Z' will be either $Z = Z_p[X]^\wedge$ or $Z^* = Z_p[X_1, \ldots, X_n]^\wedge$).

The interrelation between generic and quasi-generic extensions is as follows: Any generic extension is trivially quasi-generic. In the other direction one has:

Lemma 6.3. *If for each $1 \le m \le n$, Z_m/Z is a quasi-generic C_{p^m}-extension, and ι_m denotes the injection $C_{p^m} \to C_{p^n}$, then one obtains a generic C_{p^n}-extension Z_n^*/Z^**

as follows: Let ε_i: $Z \to Z^*$ be the morphism given by $X \mapsto X_i$ $(1 \le i \le n)$ and let Z_m' be the C_{p^m}-extension $Z^* \otimes_{\varepsilon_i} Z_m$ of Z^*. Then the Harrison product

$$\iota_1^*(Z_1')\cdot\ldots\cdot\iota_{n-1}^*(Z_{n-1}')\cdot Z_n'$$

is a generic C_{p^n}-extension of Z^*.

Proof. This is straightforward. Since we will not use the explicit form of a generic C_{p^n}-extension of Z^*, we omit the argument.

From now on, we focus on quasi-generic C_{p^n}-extensions of $Z = \mathbb{Z}_p[X]^\wedge$. Let us now prove that the extension Z_n/Z described in Prop. 6.1 is quasi-generic. Let $S \in \mathbf{C}$, $T \in \mathrm{H}(S, C_{p^n})$. Let ' denote "factor ring modulo p" throughout. Then the canonical map $\mathrm{H}(S, C_{p^n}) \longrightarrow \mathrm{H}(S', C_{p^n})$ is an isomorphism. The extension T'/S' has a description by Artin-Schreier theory:

$$T' = S'[\underline{\Theta}]/(\underline{\Theta}^p \mathbin{\dot-} \underline{\Theta} \mathbin{\dot-} (\overline{s_0},\ldots,\overline{s_{n-1}}))$$

with Θ a length n Witt vector of indeterminates, and $s_0,\ldots,s_{n-1} \in S$. On the other hand, Z_n reduces mod p to

$$Z_n' = \mathbb{F}_p[X][\underline{\Theta}]/(\underline{\Theta}^p \mathbin{\dot-} \underline{\Theta} \mathbin{\dot-} (X,0,\ldots,0)).$$

Let φ: $Z \longrightarrow S$ be defined by $\varphi(X) = s_0$. Then φ': $\mathbb{F}_p[X] \longrightarrow S'$ satisfies $\varphi'(X) = \overline{s_0}$, whence we get

$$S' \otimes_{\varphi'} Z_n' = S'[\underline{\Theta}]/(\underline{\Theta}^p \mathbin{\dot-} \underline{\Theta} \mathbin{\dot-} (\overline{s_0},0,\ldots,0)).$$

Since Φ_n is a group homomorphism (see Thm. 1.1), the two extensions T' and $S' \otimes_{\varphi'} Z_n'$ differ in the group $\mathrm{H}(S', C_{p^n})$ just by the Artin-Schreier extension $S'[\underline{\Theta}]/(\underline{\Theta}^p \mathbin{\dot-} \underline{\Theta} \mathbin{\dot-} (0,\overline{s_1},\ldots,\overline{s_{n-1}}))$. This latter extension is in the image of ι^* by Lemma 1.2. Since reduction mod p induces an isomorphism on $\mathrm{H}(-, C_{p^n})$ (as used already several times), we get the corresponding statement for T and $S \otimes_\varphi Z_n$, q.e.d.

Perhaps it has become apparent in the preceding argument that the *existence* of a quasi-generic extension Z_n/Z is just a formal consequence of Artin-Schreier theory. It is much less straightforward to give a *construction* of Z_n which does not appeal to lifting mod p. This can be done by the technique of Kummer theory plus descent employed earlier. We can put this in another perspective: It is common in algebraic number theory that one has a good description of an extension L/K by class field theory, but it is far from clear how (in the case where Kummer theory applies) L can be generated by adjunction of radicals. Let us treat just one example: Let $K = \mathbb{Q}_p(\zeta_n)$ (still ζ_n is primitive of order p^n), and L the (!) unramified cyclic field extension of degree p^n of K. (L is obtained by composing K and the unramified C_{p^n}-extension of $\mathbb{Q}_p$.) Then there exists $u \in K$ with $L = K(u^{p^{-n}})$, but which u may we take? This problem is actually related to having a good handle on (quasi-)generic extensions of $\mathbb{Z}$, since L corresponds by **0** §4 to a C_{p^n}-extension of $\mathbb{Z}_p[\zeta_n]$. Ullom (1974) has proved: Any $u \in \mathrm{U}_{2p-1}(K) - \mathrm{U}_{2p}(K)$ satisfies $L \cong K(u^{p^{-n}})$. Here $\mathrm{U}_i(K)$ is the group of principal units of level i as usual, i.e. $\mathrm{U}_i(K) = \{u \in \mathcal{O}_K \mid u \equiv 1 \ (\mathrm{mod}\ \mathfrak{p}_K^i)\}$.

Assume $p \neq 2$ for the rest of the section and write ζ for ζ_n. We shall now try to explain in which way the quasi-generic extension Z_n/Z is obtained by cyclotomic descent (Chap. I) from a Kummer extension. Nothing essentially new will be obtained; what follows should be considered as a worked example of cyclotomic descent.

Recall $R_0 = \mathbb{Z}_p[X]^\wedge[\zeta_n]$, and write R_n for the C_{p^n}-extension $Z_n[\zeta_n]$ of R_0 (cf. Prop. 6.1). We consider now $R_n[p^{-1}] \in \mathrm{H}(R_0[p^{-1}], C_{p^n})$. Then in the notation of I §1, $R_0[p^{-1}]$ is the p^n-th cyclotomic extension of $Z[p^{-1}]$. On the other hand, $R_0[p^{-1}]$ is p^n-kummerian, so we may write by 6.1 b) and 3.3:

$$R_n[p^{-1}] \cong R_0[p^{-1}]\big(p^n; b\big) \qquad \text{with } b = z^{p^n} \in R_0^*.$$

Recall the definition of Γ_n, ω, β, w, ξ from Chap. I: $\Gamma_n = \mathrm{Aut}(R_0[p^{-1}]/Z[p^{-1}]) \cong (\mathbb{Z}/p^n)^*$; $\omega\colon \Gamma_n \longrightarrow (\mathbb{Z}/p^n)^*$ the cyclotomic character (an isomorphism in the present case); β a generator of Γ_n; w a lifting of $\omega(\beta)$ to $\mathbb{Z}$ such that $w^m - 1$ is precisely divisible by the power p^n, and finally $\xi = \sum_{i=0}^{m-1} w^i \beta^{-i}$ ($m = |\Gamma_n|$; in the present case $m = p^{n-1}(p-1)$.) We proved the formula

$$(1 - w\cdot\beta^{-1})\cdot\xi \quad = \quad qp^n \qquad \text{for some } q \in \mathbb{Z},\ q \text{ prime to } p.$$

Now the fact that the C_{p^n}-extension $R_n[p^{-1}]$ of $R_0[p^{-1}]$ descends to an extension $Z_n[p^{-1}]$ of $Z[p^{-1}]$ means in the language of Thm. I 3.4: $R_n[p^{-1}] \in \mathrm{Im}(j_0)$, where j_0 is base extension: $\mathrm{H}(Z[p^{-1}], C_{p^n}) \longrightarrow \mathrm{H}(R_0[p^{-1}], C_{p^n})$. By the quoted theorem, the class of b mod p^n-th powers is in the image of exponentiation by ξ, i.e. there *must* exist an $y \in R_0[p^{-1}]$ such that

$$b \text{ is congruent to } y^\xi \text{ modulo } \big(R_0[p^{-1}]^*\big)^{p^n}.$$

Moreover it follows from the proof of I 3.4 that y can be chosen in such a way that the descent datum $\varphi = \Phi_\beta$ defining $Z_n[p^{-1}]$ is given by

$$\varphi(z) \;=\; z^w\cdot y^q \qquad (\text{and } \varphi\big|_{R_0[p^{-1}]} = \beta).$$

But the descent datum which defines $Z_n[p^{-1}]$, in other words the action of Γ_n on z, was calculated in §5, which allows to compute y^q (and y if one wants to). It is now a very good test for the theories in Chap. I and Chap. VI respectively, whether we do indeed find that $y \in R_0[p^{-1}]$. (A priori, $y \in R_n[p^{-1}]$.)

From the above formula we have $y^q = \varphi(z)/z^w = z^{\beta-w}$ in the notation of Cor. 5.7. It now follows from Cor. 5.7 that y^q is stable under C_{p^n}, hence even in R_0. Since z (and hence y^q) is congruent to 1 mod λ, one may extract the q-th root of y^q e.g. by the binomial series.

Let us sum up:

$b \in R_0 = \mathbb{Z}_p[X,\zeta]^\wedge$ is explicitly given

$$(b = \exp(p^n\eta X + (p^n\eta + p^{n-1}\eta^p)X^p + \ldots), \text{ see §5});$$

$R_n[p^{-1}]$ is the Kummer extension of $R_0[p^{-1}]$ given by adjunction of a p^n-th root z of b, and we have $\log z = \eta X + (\eta + p^{-1}\eta^p)X^p + \ldots$;

Γ_n acts on $R_n[p^{-1}]$ canonically (trivially on X, and in the natural way on $\mathbb{Z}_p[\zeta]$);

$y = \left(z^{\beta-w}\right)^{1/q}$ (with β and w as above),

and

$Z_n[p^{-1}]$ is the fixed ring of the automorphism φ on $R_n[p^{-1}]$, which equals β on $\mathbb{Z}_p[\zeta]$ and maps z to $z^w \cdot y^q$. The element y is (almost) determined by the equation

$$\log(y) = q^{-1}\cdot(\beta - w)\cdot(\eta X + (\eta + p^{-1}\eta^p)X^p + \ldots).$$

From this one finally gets that Z_n is the fixed ring of φ, considered as an automorphism of R_n = integral closure of R_0. It seems possible (not to say probable) that the formula for log(y) can be simplified further; as it stands, it is rather difficult to evaluate.

References

ARTIN, E. (1931): Über Einheiten relativ galoisscher Zahlkörper, *J. Math.* 167, p. 153–156; *Gesammelte Werke*, p. 197–200

AUSLANDER, M. and D. BUCHSBAUM (1959): On ramification theory in noetherian rings, *Amer. J. Math.* 81, p. 749–765

AUSLANDER, M. and O. GOLDMAN (1960): The Brauer group of a commutative ring, *Trans. Amer. Math. Soc.* 81, *p.* 367–409

BOREVIČ, A. (1979): Kummer extensions of rings, *J. Soviet Math.* 11, p. 514–534

BOURBAKI, N. (1961): Algèbre commutative, *Hermann*, Paris

CARTAN, H. and S. EILENBERG (1956): Homological algebra, *Princeton University Press*, Princeton, N.J.

CASSOU-NOGUES, P. and M. TAYLOR (1986): Elliptic funcions and rings of integers, *Progress in Mathematics* No. 66, *Birkhäuser*, Basel

CHASE, S. U., D. K. HARRISON, and A. ROSENBERG (1965): Galois theory and Galois cohomology of commutative rings, *Mem. Amer. Math. Soc.* No. 52 (reprinted with corrections 1968)

CHILDS, L. (1977): The group of unramified Kummer extensions of prime degree, *Proc. London Math. Soc. XXXV*, p. 407–422

—— (1984): Cyclic Stickelberger cohomology and descent of Kummer extensions, *Proc. Amer. Math. Soc.* 90, p. 505–510

COATES, J. (1977): p-adic L-functions and Iwasawa's theory, in: Algebraic number theory, *Proc. Durham Symposion*, ed. A. Fröhlich, *Academic Press*, p. 269–353

DEMEYER, F. and E. INGRAHAM (1971): Separable algebras over commutative rings, *Springer Lecture Notes in Math.* No. 181, *Springer Verlag*, Heidelberg

FALTINGS, G. (1983): Endlichkeitssätze für abelsche Varietäten über Zahlkörpern, *Invent. Math.* 73, p. 349–366

GREENBERG, R. (1973): A note on K_2 and the theory of $\mathbb{Z}_p$-extensions, *Amer. J. Math.* 100, p. 1235–1245

GREITHER, C. (1988): Cyclic Galois extensions and normal bases, *Habilitationsschrift*, Universität München

—— (1989): Unramified Kummer extensions of prime power degree, *manuscripta math.* 64, p. 261–290

—— (1989b): Generic C_{p^n}-extensions, *MPI-Report* 1989–59, Bonn

GREITHER, C. (1991): Cyclic Galois extensions and normal bases, *Trans. Amer. Math. Soc.* 326, p. 307–343

—— (1991b): Some remarks on units and normal bases, *preprint*

GREITHER, C. and R. HAGGENMÜLLER (1982): Abelsche Galoiserweiterungen von $R[X]$, *manuscr. math.* 38, p. 239–256

GREITHER, C. and R. MIRANDA (1989): Galois extensions of prime degree, *J. of Algebra* 124, p. 354–366

GROTHENDIECK, A. (1959): Techniques de descente, Séminaire Bourbaki 1959–60, Exposé No. 190, *Sécretariat Math.*, Paris

—— (FGA): Fondements de la géometrie algébrique, Séminaire Bourbaki 1957–1962, *Sécretariat Math.*, Paris

—— (1971): SGA 1. (Revêtements étales et groupe fondamental), *Springer Lecture Notes in Math.* No. 224, *Springer Verlag*, Heidelberg

GROTHENDIECK, A. and J. DIEUDONNE (EGA IV): Etude locale des schémas et des morphismes des schémas, *Publ. Math. IHES* Nos. 20, 24, 28, 32

HAGGENMÜLLER, R. (1985): Über die Gruppe der Galoiserweiterungen vom Primzahlgrad, *Habilitationsschrift*, Universität München

HARRISON, D. K. (1965): Abelian extensions of commutative rings, *Mem. Amer. Math. Soc.* No. 52 (reprinted with corrections 1968)

HARTSHORNE, R. (1976): Algebraic geometry, *Graduate Text in Math.* No. 52, *Springer Verlag*, Heidelberg

HASSE, H. (1936): Die Gruppe der p^n-primären Zahlen für einen Primteiler $\mathfrak{p}$ von p, *J. reine angew. Math.* 174, p. 174–183

—— (1949): Die Multiplikationsgruppe der abelschen Körper mit fester Galoisgruppe, *Abh. Math. Sem. Univ. Hamburg* 16, p. 29–40

IMAI, H. (1975): A remark on the rational points of abelian varieties with values in cyclotomic $\mathbb{Z}_p$-extensions, *Proc. Japan Acad.* 51, p. 12–16

IWASAWA, K. (1973): On $\mathbb{Z}_\ell$-extensions of algebraic number fields, *Ann. Math.* 98, p. 246–326

JANELIDZE G. (1982): On abelian extensions of commutative rings (Russian, with Georgian and English summary), *Bull. Acad. Sci. Georgian SSR* 108, p. 477–480

—— (1991): oral communication

JANUSZ, G. J. (1966): Separable algebras over commutative rings, *Trans. Amer. Math. Soc.* 122, p. 461–479

JENSEN, C. U. (1972): Sur les foncteurs dérivés de $\lim_{\leftarrow}$ et leurs applications en théorie des modules, *Springer Lecture Notes* 254, *Springer Verlag*, Heidelberg

KATZ, N. M. and S. LANG (1981): Finiteness theorems in geometric class field theory, *Enseign. Math.* (*II*) 27, p. 285–314

KERSTEN, I. (1983): Eine neue Kummertheorie für zyklische Galoiserweiterungen vom Grad p^2, *Algebra–Bericht* No. 45, *R. Fischer Verlag*, München

—— (1990): Brauergruppen von Körpern, *F. Viehweg – Verlag*, Braunschweig

KERSTEN, I. and J. MICHALIČEK (1985): A remark about Vandiver's conjecture, *Math. Rep. Acad. Sci. Canada VII* No. 1, p. 33–37

—— (1988): Kummer theory without roots of unity, *J. Pure Appl. Algebra* 50, p. 21–72

—— (1989): $\mathbb{Z}_p$–extensions of complex multiplication fields, *J. Number Th.* 32, p. 131–150

—— (1989b): On Vandiver's conjecture and $\mathbb{Z}_p$–extensions of $\mathbb{Q}(\zeta_{p^n})$, *J. Number Th.* 32, p. 371–386

KNUS, M.–A. and M. OJANGUREN (1974): Théorie de la descente et algèbres d'Azumaya, *Springer Lecture Notes in Math.* No. 389, *Springer Verlag*, Heidelberg

LANDSBURG, S. (1981): Patching theorems for projective modules, *J. Pure Appl. Algebra* 21, p. 261–277

LANG, S. (1959): Abelian varieties, *Interscience*, New York

—— (1983): Complex multiplication, *Grundlehren*, *Springer Verlag*, Heidelberg

LEOPOLDT, H.–W. (1962): Zur Arithmetik in abelschen Zahlkörpern, *J. reine angew. Math.* 209, p. 54–71

MAC LANE, S. (1963): Homology, *Grundlehren der math. Wiss.* 114, *Springer Verlag*, Heidelberg

MERKURJEV, A.S. (1986): On the structure of Brauer groups of fields, *Math. USSR* (*Izvestiya*) 27 (1), p. 141–157

MIKI, H. (1974): On $\mathbb{Z}_p$–extensions of complete p–adic power series fields and function fields, *J. Fac. Sci. Univ. Tokyo Sect. IA Math.* 21, p. 377–393

MILNE, J. S. (1986): Arithmetic duality theorems, *Perspectives in Math.*, *Academic Press*, Boston

—— (1986b): Abelian varieties, in: Arithmetic geometry, ed. G. Cornell and J. Silverman, *Springer Verlag*, Heidelberg

MUMFORD, D. (1964): Lectures on curves on an algebraic surface, *Harvard University Press*, Cambridge (Massachusetts)

NAGAHARA, T. and A. NAKAJIMA (1971): On cyclic extensions of commutative rings, *Math. J. Okayama Univ.* 15, p. 81–90

NEUKIRCH, J. (1969): Klassenkörpertheorie, *Bibliographisches Institut*, Mannheim

RIBET, K. (1980): Division fields of abelian varieties with complex multiplication, *Mem. Soc. Math. France (2ème serie)* No. 2, p. 75–94

—— (1981): Torsion points of abelian varieties, Appendix to: Katz and Lang (1981), p. 315–319

RIEHM, C. (1970): The corestriction of algebraic structures, *Inv. Math.* 11, p. 73–98

ROSENLICHT, M. (1961): Toroidal algebraic groups, *Proc. Amer. Math. Soc.* 12, p. 984–988

SALTMAN, D. (1982): Generic Galois extensions and problems in field theory, *Advances Math.* 43, p. 250–283

SCHAPPACHER, N. (1977): Zur Existenz einfacher abelscher Varietäten mit komplexer Multiplikation, *J. reine angew. Math.* 292, p. 186–190

SCHMIDT, C.-G. (1984): Arithmetik abelscher Varietäten mit komplexer Multiplikation, *Springer Lecture Notes in Math.* No. 1082, *Springer Verlag*, Heidelberg

SERRE, J.-P. (1964): Cohomologie galoisienne, *Springer Lecture Notes in Math.* No. 5, *Springer Verlag*, Heidelberg

—— (1968): Corps locaux, *Hermann*, Paris

—— (1974): Letters to B. Mazur, January 1974

—— (1978): Sur le résidu de la fonction zêta p-adique d'un corps de nombres, *C. R. Acad. Sci. Paris* 287, p. 183–188

ULLOM, S. (1974): Integral representations afforded by ambiguous ideals in some abelian extensions, *J. Number Th.* 6, p. 32–49

WANG, S. (1948): A counterexample to Grunwald's theorem, *Ann. Math.* (2) 49, p. 1008–1009

WASHINGTON, L. (1982): Introduction to cyclotomic fields, *Graduate Text* No. 83, *Springer Verlag*, Heidelberg

WATERHOUSE, W. (1987): A unified Kummer–Artin–Schreier sequence, *Math. Ann.* 277, p. 447–451

WENNINGER, C.-H. (1991): Corestriction of Galois algebras, *Journal of Algebra* 144, p. 359–370

WITT, E. (1936): Zyklische Körper und Algberen der Charakteristik p vom Grad p^n, *J. reine angew. Math.* 174, p. 126–140

WYLER, T. (1987): Torsors under abelian p-groups, *J. Pure Appl. Algebra* 45, p. 273–286

INDEX